AF760434

Springer Series in

CHEMICAL PHYSICS 103

Springer Series in
CHEMICAL PHYSICS

Series Editors: A. W. Castleman, Jr. J. P. Toennies K. Yamanouchi W. Zinth

The purpose of this series is to provide comprehensive up-to-date monographs in both well established disciplines and emerging research areas within the broad fields of chemical physics and physical chemistry. The books deal with both fundamental science and applications, and may have either a theoretical or an experimental emphasis. They are aimed primarily at researchers and graduate students in chemical physics and related fields.

Please view available titles in *Springer Series in Chemical Physics* on series homepage http://www.springer.com/series/676

Kaoru Yamanouchi
Mauro Nisoli
Wendell T. Hill, III
Editors

Progress in Ultrafast Intense Laser Science

Volume VIII

With 115 Figures

Springer

Editors

Professor Kaoru Yamanouchi
Department of Chemistry
The Universtity of Tokyo
Tokyo
Japan

Professor Wendell T. Hill, III
IPST
University of Maryland
MD
USA

Professor Mauro Nisoli
Department of Physics
Politecnico di Milano
Milano
Italy

Series Editors:

Professor A.W. Castleman, Jr.
Department of Chemistry, The Pennsylvania State University
PA, USA

Professor J.P. Toennies
Max-Planck-Institut für Strömungsforschung
Göttingen, Germany

Professor K. Yamanouchi
Department of Chemistry, The University of Tokyo
7-3-1 Hongo, Bunkyo-ku, Tokyo 113-0033, Japan

Professor W. Zinth
Universität München, Institut für Medizinische Optik
München, Germany

Springer Series in Chemical Physics ISSN 0172-6218
ISBN 978-3-642-28725-1 ISBN 978-3-642-28726-8 (eBook)
DOI 10.1007/978-3-642-28726-8
Springer Heidelberg New York Dordrecht London

Library of Congress Control Number: 2006927806

© Springer-Verlag Berlin Heidelberg 2012
This work is subject to copyright. All rights are reserved by the Publisher, whether the whole or part of the material is concerned, specifically the rights of translation, reprinting, reuse of illustrations, recitation, broadcasting, reproduction on microfilms or in any other physical way, and transmission or information storage and retrieval, electronic adaptation, computer software, or by similar or dissimilar methodology now known or hereafter developed. Exempted from this legal reservation are brief excerpts in connection with reviews or scholarly analysis or material supplied specifically for the purpose of being entered and executed on a computer system, for exclusive use by the purchaser of the work. Duplication of this publication or parts thereof is permitted only under the provisions of the Copyright Law of the Publisher's location, in its current version, and permission for use must always be obtained from Springer. Permissions for use may be obtained through RightsLink at the Copyright Clearance Center. Violations are liable to prosecution under the respective Copyright Law.
The use of general descriptive names, registered names, trademarks, service marks, etc. in this publication does not imply, even in the absence of a specific statement, that such names are exempt from the relevant protective laws and regulations and therefore free for general use.
While the advice and information in this book are believed to be true and accurate at the date of publication, neither the authors nor the editors nor the publisher can accept any legal responsibility for any errors or omissions that may be made. The publisher makes no warranty, express or implied, with respect to the material contained herein.

Printed on acid-free paper

Springer is part of Springer Science+Business Media (www.springer.com)

Preface

We are pleased to present the eighth volume of Progress in Ultrafast Intense Laser Science. As the frontiers of ultrafast intense laser science rapidly expand ever outward, there continues to be a growing demand for an introduction to this interdisciplinary research field that is at once widely accessible and capable of delivering cutting-edge developments. Our series aims to respond to this call by providing a compilation of concise review-style articles written by researchers at the forefront of this research field, so that researchers with different backgrounds as well as graduate students can easily grasp the essential aspects.

As in previous volumes of PUILS, each chapter of this book begins with an introductory part, in which a clear and concise overview of the topic and its significance is given, and moves onto a description of the authors' most recent research results. All the chapters are peer-reviewed. The chapters of this eighth volume cover a diverse range of the interdisciplinary research field, and the topics may be grouped into three categories: molecules interacting with ultrashort and intense laser fields (Chaps. 1–4), advanced technologies for the characterization of ultrashort laser pulses and their applications (Chaps. 5 and 6), and laser plasma formation and laser acceleration (Chaps. 7–9).

From the third volume, the PUILS series has been edited in liaison with the activities of Center for Ultrafast Intense Laser Science in the University of Tokyo, which has also been responsible for sponsoring the series and making the regular publication of its volumes possible. From the fifth volume, the Consortium on Education and Research on Advanced Laser Science, the University of Tokyo, has joined this publication activity as one of the sponsoring programs. The series, designed to stimulate interdisciplinary discussion at the forefront of ultrafast intense laser science, has also collaborated since its inception with the annual symposium series of ISUILS (http://www.isuils.jp), sponsored by JILS (Japan Intense Light Field Science Society).

We would like to take this opportunity to thank all the authors who have kindly contributed to the PUILS series by describing their most recent work at the frontiers of ultrafast intense laser science. We also thank the reviewers who have read the submitted manuscripts carefully. One of the coeditors (KY) thanks Ms. Chie Sakuta for her help with the editing processes. Last but not least, our gratitude goes out to Dr. Claus Ascheron, Physics Editor of Springer-Verlag at Heidelberg, for his kind support.

We hope this volume will convey the excitement of ultrafast intense laser science to the readers and stimulate interdisciplinary interactions among researchers, thus paving the way to explorations of new frontiers.

Tokyo	Kaoru Yamanouchi
Milano	Mauro Nisoli
Maryland	Wendell T. Hill, III

Contents

Chapter 1
Probing Electron Dynamics in Simple Molecules with Attosecond Pulses

Paula Rivière, Alicia Palacios, Jhon Fredy Pérez-Torres, and Fernando Martín

Abstract Attosecond pulses are an ideal tool to explore electron and nuclear dynamics in atoms and molecules. Either as single attosecond pulses (SAP), in attosecond pulse trains (APT), or in combination with infrared (IR) pulses, these pulses, with frequencies in the VUV-XUV regime, have been widely used to probe ionization, electron tunneling, or autoionization in atoms. More recently, similar processes have been studied in molecules. A correct theoretical description of such processes in molecules often requires a fully dimensional treatment due to the important role of nuclear motion and electron correlation. This restricts ab initio calculations to the simplest molecules. In this chapter, we discuss single ionization of hydrogen molecules (H_2 and D_2) induced by time-delayed SAP+IR and APT+IR schemes. Ab initio time-dependent theoretical calculations are compared with existing experiments.

1.1 Introduction

Attosecond laser pulses have become a standard tool to explore electron and nuclear dynamics in atoms and molecules. While nuclear dynamics in molecules takes place in the time range of few femtoseconds (the vibrational period of H_2^+ is 15 fs), and can therefore be probed with IR laser pulses, electron dynamics is much faster: the Bohr orbit time, or period of the 1s orbital of the hydrogen atom, is

---

P. Rivière (✉) · A. Palacios · J.F. Pérez-Torres
Departamento de Química, Módulo 13, Universidad Autónoma de Madrid, 28049, Madrid, Spain
e-mail: paula.riviere@uam.es; alicia.palacios@uam.es; jhon.perez@uam.es

F. Martín
Departamento de Química, Módulo 13, Universidad Autónoma de Madrid, 28049, Madrid, Spain

Instituto Madrileño de Estudios Avanzados en Nanociencia, Cantoblanco, 28049 Madrid, Spain
e-mail: fernando.martin@uam.es

K. Yamanouchi et al. (eds.), *Progress in Ultrafast Intense Laser Science VIII*,
Springer Series in Chemical Physics 103, DOI 10.1007/978-3-642-28726-8_1,
© Springer-Verlag Berlin Heidelberg 2012

$\sim$150 as. This makes attosecond pulses optimal to explore electron dynamics in, for instance, autoionization where both electrons and nuclei move in comparable time scales. These ultrashort pulses are typically generated by high-order harmonic generation (HHG) from atoms and molecules, which produces typical wavelengths in the vacuum ultraviolet (VUV) and soft X-ray (XUV) energy regions, with pulse durations of even less than 100 as [1]. The frequencies and duration of these pulsed radiation sources make them well suited to induce electronic excitation and ionization processes in atoms and molecules. Due to the nature of the generation process, the intensity of these pulses is still far from the intensities reached for longer pulses. However, a progressive increase in intensities ($10^9 \rightarrow 10^{14}$ W/cm^2) has led to the possibility of exploring nonlinear processes as well [2]. Free-electron lasers (FEL) and forthcoming XFEL facilities will also produce intense pulses with durations of the order of femtoseconds in the XUV and X-ray photon energy range [3,4].

The obvious advantage of attosecond pulses is that, upon one-photon absorption in the VUV-XUV frequency domain, they can launch and probe electron and nuclear wave packets (NWPs) with high temporal resolution. This enables real-time observations of dynamical processes in atoms, as it has been shown in studies on electron tunneling [5] or exponential autoionization (Auger) decay of core-excited states whose lifetimes lie in the femtosecond range [6]. In molecules, attosecond pulses can be used to study a wide range of phenomena involving electron dynamics, such as tomographic imaging of molecular orbitals [7], beating in coherent sums of states [8, 9], or interferences due to photoelectron emission from different atomic centers [10].

Interestingly, attosecond electron dynamics in molecules can also be explored using long pulses and strong fields. Some examples are the use of electron pulses for probing sub-fs electron dynamics in H_2 [11, 12] and for controlling electron localization in molecular dissociation of H_2^+ and D_2^+ [13–15]. These intense and relatively longer pulses, with durations of tens of fs, are suitable to probe nuclear dynamics [16, 17] or its correlation with electron localization [18].

The interpretation and analysis of the complex dynamical processes induced in molecules by ultrashort pulses very often require support from nearly exact theoretical methods. Such methods are available for the one-electron molecular hydrogen ion [19–23] and have been applied to interpret several experimental works [13–15, 24] that have explored the combined electron and nuclear dynamics. The neutral H_2 and D_2 molecules show a richer dynamics due to the presence of electron correlation. Until very recently, the complexity introduced by the second electron has made calculations on these molecules computationally prohibitive. Approximate theoretical models have shed some light on electron correlation in hydrogen molecules [24,25], but, as we will see below, an appropriate description of these systems requires a full-dimensional treatment that accounts for all electronic and nuclear degrees of freedom.

For example, in XUV photoionization of H_2, photons from the XUV laser pulse can populate doubly excited states, which may either lead to neutral dissociation

($H_2^{**} \rightarrow$ H+H) or decay via electron correlation to the H_2^+ dissociative and non-dissociative continua, in an autoionizing process ($H_2^{**} \rightarrow H_2^+ + e^-$) [26]. H_2 autoionization is observable in the kinetic energy release (KER) differential probabilities of the fragments as well as in the fully differential electron angular distributions. The lifetime of autoionizing states in this system is in the range of a few femtoseconds (i.e., for the Q_1 series of autoionizing states), which occurs in the same time scale of vibrations and dissociation in the molecule. Therefore, the role of nuclear dynamics cannot be neglected, and the Franck–Condon approximation is no longer valid.

The appearance of FEL in the XUV and X-ray regions, together with new sources of HHG, has increased considerably the number of applications of two-color fields with XUV and strong IR pulses. As we have seen, intense IR fields or attosecond pulses can be used separately to probe atomic and molecular dynamics; however, the combination of both is specially fruitful, since it combines two time ranges (atto-femtoseconds) and two ranges of frequencies. The theoretical description of ionization and dissociation of H_2 and D_2 arising in these two-color schemes is the main topic of this chapter.

The chapter is organized in four sections. A review on the different setups for probing atoms and molecules with ultrashort pulses is given in Sect. 1.2. Our time-dependent theoretical approach is discussed in Sect. 1.3. Two specific problems will be discussed in detail in Sect. 1.4: single ionization of hydrogenic molecules using IR pulses combined with (1) single attosecond pulses (SAP), SAP+IR, and (2) attosecond pulse trains (APT), APT+IR. Conclusions are drawn in Sect. 1.5.

1.2 Probing Electron Dynamics with Ultrashort Pulses

Ultrashort electron and nuclear dynamics in atoms and molecules has been probed either with intense IR pulses, with attosecond UV pulses or with a combination of them, depending on the process to be induced and traced. This gives rise to a large variety of pump-probe schemes. While research in atoms has been extensive, both experimentally and theoretically (for reviews on the field, see [27–31] and references therein), the study of attosecond dynamics in molecules is more scarce due to its complexity. In some cases, the nuclear motion is entangled and affects the electronic dynamics, leading to phenomena that do not exist in atomic targets and are thus unpredictable by using simple models or theoretical methods based on the fixed-nuclei approximation (FNA).

In addition, the validity of different theoretical approaches is linked to the properties of the light source. The interaction of atoms and molecules with intense laser radiation has been traditionally classified in two regimes. In the so-called tunneling regime, the electromagnetic field at a given intensity and frequency is strong and slow enough to tilt the potential that keeps the electron bound to the nucleus allowing tunnel ionization through the resulting potential barrier.

Such process has been widely described using theoretical approaches in a *dressed states* picture [32]. On the other hand, one can increase the field frequency such that the variation of the field with time is so sudden that the electron has not enough time to tunnel. This is the multiphoton regime, which can frequently be described in the framework of perturbation theory. In brief, higher intensities and lower frequencies of the electromagnetic field favor ionization through a tunneling process, while lower intensities and higher frequencies favor ionization through single- and multiphoton absorption. A simple quantitative picture on the separation of ionization regimes is given through the Keldysh formalism [33], by relating the atomic (molecular) ionization potential, I_p, with the ponderomotive energy U_p, which is the maximum energy that the electron absorbs from the field during its excursion in the continuum. The Keldysh parameter is defined as $\gamma = \sqrt{I_p/2U_p}$, where $U_p = I/4\omega^2$, with I being the laser intensity and ω being its frequency. For values of $\gamma \ll 1$, ionization takes place through the tunneling process, whereas for $\gamma > 1$, the interaction is described within the multiphoton regime.

In the following, we summarize recent experimental studies on ionization induced by pulsed radiation. They are classified according to the different light sources and pump-probe schemes employed. We will refer to XUV frequencies, but the same conclusions can be applied to EUV or VUV (extreme or vacuum ultraviolet light):

1.2.1 Intense IR Fields

Although strong IR femtosecond laser pulses have been widely used to study femtosecond dynamics in atoms and molecules, only in the last decade they have been applied to explore attosecond dynamics (i.e., see [11, 13–16, 18]). Even with the long period of IR pulses ($T \sim 2.7$ fs for $\lambda = 800$ nm), subfemtosecond dynamics can be studied. Several works have explored different processes observed in H_2^+ ionization: electron localization during dissociative ionization [13, 14] and its dependence with the carrier-envelope phase (CEP) of the IR pulse, which is the relative phase between the light oscillations and the time-dependent envelope of the finite pulse, diffraction effects [34], or purely subfemtosecond effects such as the ejection of multiple bursts during ionization [35].

1.2.2 XUV Fields

For these relatively high frequencies, we already mentioned that photoabsorption events become perturbative. Since the effect of the laser in the atom or molecule can be reduced up to a good approximation to the absorption or emission of a single photon, the process can be accurately described using first-order perturbation theory. Currently, the VUV or XUV frequencies of attosecond pulses experimentally

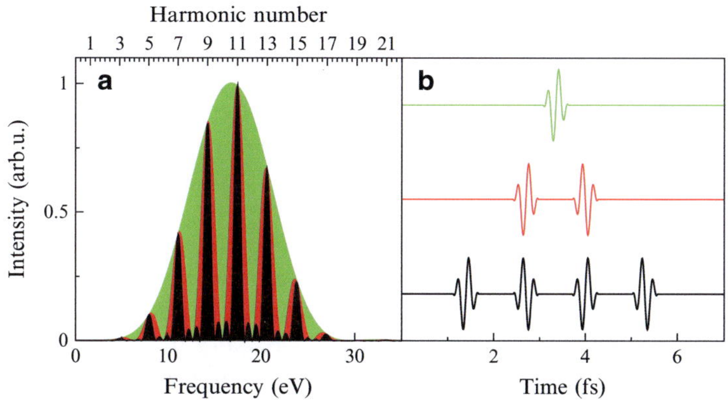

Fig. 1.1 Energy spectrum (**a**) as a function of the number of pulses in (**b**), for attosecond pulses of FWHM∼200 as and $\lambda \simeq 87$ nm (9th harmonic of a 780-nm infrared wavelength). The delay between the pulses is $T/2$, where T is the IR frequency, so there are two atto pulses per IR cycle

available and obtained from HHG lie roughly from the 9th to the 27th harmonic of the IR generating frequency, in the 30–90 nm range. They are typically able to excite or ionize the electrons of atoms and molecules, with relatively weak intensities mostly related to one-photon, perturbative processes.

In experiments using ultrashort VUV-XUV pulses, there are two common scenarios: either SAP, which have a very broad energy bandwidth of several eV, or APT, whose energy spectrum is formed by several spikes of narrower width (∼1 eV). Figure 1.1 shows the energy spectra for XUV pulsed radiation containing a variable number of attosecond pulses in the train.

Notice that for pulse trains containing two atto pulses per IR cycle, the frequency spectrum shows odd harmonics of the generating frequency. However, for electromagnetic fields with one atto pulse per IR cycle, frequencies of both even and odd harmonics appear. The latter scheme can be achieved by using two-color IR fields to generate the APT [36]. Characterization of APT can be performed by measuring the relative phases between different harmonics [37, 38]. In the last years, an increased number of applications of XUV pump-probe schemes have been developed: for instance, in studies of two-photon atomic photoionization [39, 40] or of electron motion in dissociating H_2^+ [9].

1.2.3 SAP + IR

The combination of XUV and IR pulses presents several advantages compared with one-color schemes. In these setups, the notion of time delay between the pulses is meaningful as long as the CEP of the IR is fixed [41]. Besides, in those cases where there is a large difference in intensity and frequency between the XUV pump and the IR probe, the strong field approximation (SFA) is suitable to describe single ionization processes in atoms [33,42]. In the SFA, the target is initially in its ground state, where the electron is considered to be isolated from the IR field, but can absorb an XUV photon. The resulting ionized electron is only driven by the IR field; thus, it is supposed to be isolated from the ionic potential [43]. The latter steering process, in which the IR deflects the electron and changes its momentum and angular distribution, is known as *streaking* [44,45], and it strongly depends on the IR phase at the moment of ionization.

Despite the success of the SFA in the description of nonperturbative interactions of intense lasers with atoms, a complete characterization of ionization processes can only be performed by including electron correlation effects. In particular, full-dimensional calculations eventually become mandatory for molecular targets due to the relevance of nuclear motion in multiphoton ionization processes [46, 47]. This has been shown in experimental and theoretical works on electron localization control in H_2^+ dissociative ionization [13,23]: an initial XUV pulse is used to excite the molecular ion into the $2p\sigma_u$ dissociative state. The IR pulse is then used to steer the electron between the nuclei. As the molecule dissociates, it reaches an internuclear distance for which the absorption of an IR photon is no longer enough to overcome the potential barrier between the nuclei (i.e., the barrier reaches the energy of the $2p\sigma_u$ state). The electron is then trapped in one of the nuclei. By switching the time delay between the XUV and IR pulses (or, equivalently for long IR pulses, the CEP of the IR pulse for a fixed time delay), the localization of the electron in one of the nuclei can be controlled with high efficiency, up to 85% [23]. SAP+IR schemes have also been used to perform electron tunneling spectroscopy in molecular dissociation [48].

The notion of *dressed states* mentioned above is useful in this context. This term refers to the effect of the strong IR field on the electronic states of the atom or molecule. In a Born-Oppenheimer (BO) picture, in which we separate electronic motion from nuclear motion, we could draw new dressed-BO curves by summing or subtracting any number of IR photons. The different curves will be energetically accessible or not depending on the number of photons absorbed, i.e., on the IR intensity. Avoided crossings appear that give rise to a decrease of the dissociation barrier in the molecule. This effect is called *bond softening* (see, e.g., [17,49]). In this process, the $1s\sigma_g$ and $2p\sigma_u$ states, which are degenerate at $R \to \infty$ in the field-free case, are lifted downward (the lower state) and upward (the upper one) due to the IR field (see Fig. 1.2). As a result, vibrational states of the $1s\sigma_g$ that would be vibrationally bound in the field-free case can dissociate, leading to ionic fragments with low kinetic energy. The corresponding increase of the barrier in the upper level is called *bond hardening*.

1.2.4 APT+IR

APT in combination with finite IR pulses allow for excitation and ionization by absorption of photons with the frequencies of different harmonics from the APT, together with the emission/absorption of one or a few IR photons. If the pulse

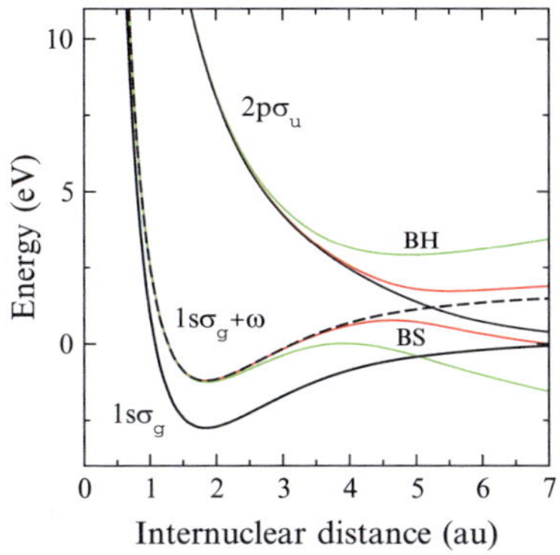

Fig. 1.2 Effect of an intense 800-nm electric field on the H_2^+ levels. The diabatic states $1s\sigma_g$ and $2p\sigma_u$ are shown with black lines, together with the $1s\sigma_g$ after absorption of one photon (*dashed black*). The adiabatic potential energy curves for $I = 10^{13}$ W/cm^2 (*red*) and 10^{14} W/cm^2 (*green*) are also shown. The regions of vibrational trapping or bond hardening (BH) and bond softening (BS) are pointed out

duration of the IR is long enough, the photon absorption of adjacent harmonics plus or minus an IR photon leads to interferences, as a result of the same final state being reached through different paths. This is the underlying physics behind the RABBITT experimental technique, used for the characterization of APT [37, 38]. Similarly, for strong IR fields, one reaches the streaking regime, which has led to interesting experimental applications as in the characterization of SAP [41,50] or in the quantum stroboscope [51].

The APT+IR combination was first used to probe electron dynamics in atoms, showing a strong variation of ionization yields with the relative times of the pulses [52]. The use of these schemes for electron localization in molecules requires only one attosecond pulse per IR cycle, such that the electronic wave packet (EWP) is always driven along the same direction of the polarization axis [53]. The observed photoelectron spectra are strongly dependent on the existence of resonant absorption paths [54]. Its complex dependence on the APT-IR time delay has been explained in terms of interferences between electron wave packets [55] or within Floquet states pictures [32]. In this process, the IR field accelerates the ejected electron, which is then localized in one of the nuclei, whereas the interference between electron wave packets generated by different pulses in the APT enhances/suppresses the localization probability [23]. The pulse parameters of the APT and IR pulses can be chosen such that the electron swaps between the nuclei after localization has taken place [56].

Nowadays, new experimental APT+IR techniques, already available and potentially powerful, combine APT with two-color IR pulses: these are (APT+IR)+IR schemes, where the APT and the first IR pulse are phase-locked. The second IR pulse, with variable intensity, is used as a probe. These schemes enable the possibility of probing optically forbidden states, with the same symmetry as the molecular ground state.

In most pump-probe schemes used in molecules, either with SAP or with APTs, the XUV pulses are weak, and the IR ones, though more intense than the XUV pulses, still keep the ionization rate within the multiphoton regime. For example, for IR fields with a typical wavelength of 800 nm and intensity 10^{13} W/cm^2, the Keldysh parameter is $\gamma = 3.7 \gg 1$ (where the ionization potential of the H_2 molecule at the equilibrium distance is $I_p = -0.597$ a.u.). In this situation, the XUV pulses can be considered to ionize or excite the electrons in a one-photon perturbative process, while the effect of the IR is to accelerate the electrons in the continuum by absorption/emission of a few photons. The rest of the chapter will focus on recent work performed in this context.

1.3 Theoretical Method

In this section, we present a brief summary of a recently proposed time-dependent spectral method to investigate ionization and dissociation of H_2 and D_2 by using the SAP+IR and APT+IR schemes mentioned above. The method accounts for all vibrational and electronic degrees of freedom and accurately describes electron correlation except the overall rotation of the molecule.

1.3.1 Time-Dependent Schrödinger Equation

The time-dependent Schrödinger equation is written as

$$\left[\hat{H}_0(\mathbf{r}, R) + \hat{V}(t) - \mathrm{i}\frac{\partial}{\partial t}\right]\Phi(\mathbf{r}, R, t) = 0, \tag{1.1}$$

where $\mathbf{r}$ represents the electronic coordinates $\mathbf{r}_1$ and $\mathbf{r}_2$ and R is the internuclear distance. $\hat{H}_0$ is the H_2 field-free nonrelativistic Hamiltonian

$$\hat{H}_0(\mathbf{r}, R) = \hat{T}(R) + \hat{H}_{\mathrm{el}}(\mathbf{r}, R), \tag{1.2}$$

where $\hat{T}(R) = -\frac{\hat{\nabla}_R^2}{2\mu}$ is the relative kinetic energy of the nuclei, μ is the reduced mass, and $\hat{H}_{\mathrm{el}}(\mathbf{r}, R)$ is the electronic Hamiltonian which includes the $1/R$ repulsion term. The interaction potential between the laser field and the H_2 molecule, $\hat{V}(t)$, is obtained within the dipole approximation, and using the velocity gauge, it can be expressed as

$$\hat{V}(t) = \hat{\mathbf{p}} \cdot \hat{\mathbf{A}}(t), \tag{1.3}$$

where $\hat{\mathbf{p}}$ is the momentum operator of the electron and $\hat{\mathbf{A}}(t)$ is the vector potential of the electromagnetic field, which is described classically. The finite pulse duration is usually introduced through time-dependent envelopes. It should be noted that although most theoretical work is done with these ideal (Gaussian or sine-/cosine-like envelopes) pulses, experimental pulses usually contain time-varying frequencies (i.e., chirped pulses), pedestals, and complex spectral profiles. For example, it has been shown that for relatively short pulses (few fs), chirps and spectral shapes strongly influence photoelectron angular distributions in H_2 dissociative ionization [57] and pedestals strongly affect nuclear dynamics in D_2^+ [17].

1.3.2 Molecular Electronic and Nuclear States

The time-dependent Schrödinger equation is solved by expanding $\Phi(\mathbf{r}, R, t)$ in a basis of molecular states. Evaluation of these states in H_2 and D_2 requires the use of specific methods described below.

A well-known complication in the evaluation of the electronic continuum states of many-electron atoms and molecules is the presence of Feshbach resonances, corresponding to multiply excited electronic states. These are quasi-bound states embedded and coupled to the ionization continuum [58, 59], and therefore, they are autoionizing states that have a finite lifetime before they decay into the continuum. Their properties are fully determined by electron correlation, which plays a more important role here than for bound states. A resonant state is characterized by its energy position (its potential energy curve for molecules) and its autoionization width, although this information does not provide a complete description of the dynamics of autoionization in molecular systems [60].

In molecules, the conventional Fano picture that describes atomic autoionization is modified, since the lifetimes of molecular autoionizing states are of the same order as the typical time required by the nuclei to move and, in addition, they vary with internuclear distance. Indeed, the population of the Feshbach resonances can lead to the formation of dissociative and non-dissociative molecular ionic states but also to the dissociation of the molecule into neutral fragments, due to the repulsive character of the potential energy curves associated with autoionizing states. These processes occur simultaneously, leading to complicated interference phenomena [61–63]. Therefore, a wave function which is a simple product of an electronic function times a nuclear function is not valid to describe all possible ionization and dissociation pathways.

In order to properly describe autoionization in the hydrogen molecule, we use the usual Feshbach procedure [59] in which the nonresonant and resonant parts of the wave function are assigned to two orthogonal complementary subspaces P and $Q = 1 - P$, respectively. Then $\Phi(\mathbf{r}, R, t)$ is expanded in the basis of molecular eigenstates associated with each subspace. Since these states are not eigenfunctions of the total field-free Hamiltonian, they will be coupled during the time evolution (see Sect. 1.3.3).

In the following, we will assume that nonadiabatic couplings are negligible (BO approximation). The eigenstates associated with the electronic continuum in P

subspace are written as $\{\psi_\alpha^{\epsilon l}(\mathbf{r}, R)\chi_{v_\alpha}(R)\}$, where the electronic wave function $\psi_\alpha^{\epsilon l}$ satisfies the projected equation for nonresonant electron scattering

$$\left[\hat{P}\hat{H}_{\rm el}\hat{P} - \mathcal{E}(R)\right]\psi_\alpha^{\epsilon l}(\mathbf{r}, R) = 0, \quad \mathcal{E}(R) = E_\alpha(R) + \epsilon, \tag{1.4}$$

where α denotes the full set of quantum numbers for the electronic state of the residual molecular ion H_2^+ with BO energy $E_\alpha(R)$ at the internuclear distance R. The indices l and ϵ denote, respectively, the angular momentum and the kinetic energy of the ejected electron. The nuclear wave function χ_{v_α} satisfies the equation

$$\left[\hat{T}(R) + \mathcal{E}(R) - W_{v_\alpha}\right]\chi_{v_\alpha}(R) = 0, \tag{1.5}$$

where W_{v_α} is the vibronic energy (electronic plus vibrational) in channel α. The eigenstates associated with the Q subspace are written as $\{\phi_r(\mathbf{r}, R)\Upsilon_{k_r}(R)\}$, where ϕ_r is the electronic wave function of the r autoionizing state that satisfies the projected Schrödinger equation

$$\left[\hat{Q}\hat{H}_{\rm el}\hat{Q} - \mathcal{E}_r(R)\right]\phi_r(\mathbf{r}, R) = 0, \tag{1.6}$$

where $\mathcal{E}_r(R)$ denotes the BO energy of the resonance r at the internuclear distance R, and the nuclear wave function Υ_{k_r} satisfies

$$\left[\hat{T}(R) + \mathcal{E}_r(R) - W_{k_r}\right]\Upsilon_{k_r}(R) = 0, \tag{1.7}$$

in which W_{k_r} corresponds to the vibronic energy of the resonant states r. The bound states, including the initial ground state, of H_2 are written as $\{\psi_i(\mathbf{r}, R)\chi_{v_i}(R)\}$, where ψ_i is an eigenstate of the unprojected electronic Hamiltonian $\hat{H}_{\rm el}$, and χ_{v_i} is the corresponding BO vibrational wave function. The corresponding vibronic energy is W_{v_i}.

1.3.3 Time-Dependent Close-Coupling Method

The time dependent wave function $\Phi(\mathbf{r}, R, t)$ can now be expanded in the basis of the molecular eigenstates, previously obtained in the Feshbach formalism, as

$$\begin{aligned}\Phi(\mathbf{r}, R, t) = &\sum_i \sum\!\!\!\!\!\!\int_{v_i} C_{iv_i}(t)\psi_i(\mathbf{r}, R)\chi_{v_i}(R)e^{-iW_{iv_i}t} \\ &+\sum_r \sum\!\!\!\!\!\!\int_{k_r} C_{rk_r}(t)\phi_r(\mathbf{r}, R)\Upsilon_{k_r}(R)e^{-iW_{k_r}t} \\ &+\sum_{\alpha l}\int d\epsilon \sum\!\!\!\!\!\!\int_{v_\alpha} C_{\alpha v_\alpha}^{\epsilon l}(t)\psi_\alpha^{\epsilon l}(\mathbf{r}, R)\chi_{v_\alpha}(R)e^{-iW_{v_\alpha}t}.\end{aligned} \tag{1.8}$$

Atomic units are used throughout the text, unless otherwise stated ($\hbar = m = e = 1$, where m and e are the electronic mass and charge, respectively). Introducing this ansatz into the TDSE and projecting into the basis states, a system of coupled differential equations is obtained. Since P and Q are eigenstates of different Hamiltonians ($\hat{P}\hat{H}_{\text{el}}\hat{P}$ and $\hat{Q}\hat{H}_{\text{el}}\hat{Q}$, respectively), they will be coupled by the exact $\hat{H}_{\text{el}}$ Hamiltonian, leading to the so-called electrostatic couplings. These $P - Q$ couplings account for autoionization from the ϕ_r doubly excited states to the nonresonant continuum states $\psi_\alpha^{\epsilon l}$. The matrix representation of the set of coupled differential equations in the interaction picture is written as

$$i\begin{pmatrix} \dot{C}_{iv_i}(t) \\ \dot{C}_{rk_r}(t) \\ \dot{C}^{\epsilon l}_{\alpha v_\alpha}(t) \end{pmatrix}
= \begin{pmatrix} \hat{A}(t)p_{iv_i,iv_i} & \hat{A}(t)p_{iv_i,rk_r} & \hat{A}(t)p^{\epsilon l}_{iv_i,\alpha v_\alpha} \\ \hat{A}(t)p_{rk_r,ivi} & \hat{A}(t)p_{rk_r,rk_r} & (\hat{Q}\hat{H}\hat{P} + \hat{A}(t)p)^{\epsilon l}_{rk_r,\alpha v_\alpha} \\ \hat{A}(t)p^{\epsilon l}_{\alpha v_\alpha,i v_i} & (\hat{P}\hat{H}\hat{Q} + \hat{A}(t)p)^{\epsilon l}_{\alpha v_\alpha,rk_r} & \hat{A}(t)p_{\alpha v_\alpha,\alpha v_\alpha} \end{pmatrix} \begin{pmatrix} C_{iv_i}(t) \\ C_{rk_r}(t) \\ C^{\epsilon l}_{\alpha v_\alpha}(t), \end{pmatrix} \tag{1.9}$$

where $p_{i v_i,\text{jk}_j}$ stands for the dipole matrix element between the i, v_i and the j, k_j vibronic states and the $C_{iv_i}(t)$, $C_{rk_r}(t)$, and $C^{\epsilon l}_{\alpha v_\alpha}(t)$ are the time-dependent coefficients, written in the interaction picture, that are associated with the bound states, the electronic continua and the doubly excited states, respectively. Notice that, in the above equation, non-adiabatic couplings have been neglected. These couplings usually play a minor role in ionization processes except when very slow electrons are ejected or narrow avoided crossings between different autoionizing states are present in the region under investigation. This set of coupled differential equations (1.9) is integrated in time by imposing the appropriate initial condition, e.g., $C_{i,v_i=0}(t=0)=1$ when the molecule is in its ground state $i=g$. The integration must be carried out for values of time larger than the pulse duration plus the lifetime of the Q_n resonant states in order to ensure that autoionization has occurred. The amplitudes for different channels are extracted through projection into the eigenstates of the unperturbed Hamiltonian for each subspace. Given that those are the basis states in which the wave packet is expanded, the time-dependent coefficients are already the amplitudes for each vibronic state, and any dissociative photoionization probability can be accurately extracted with the appropriate linear combination of those. This is one of the main advantages of the method.

1.3.4 Asymmetric Electron Ejection

1.3.4.1 Laboratory-Frame vs Molecular-Frame

Dissociative ionization of H_2 may lend itself to the observation of both laboratory-frame and molecular-frame asymmetries in the electron ejection. The first corresponds to an asymmetry in the fragment ejection along the laser polarization axis, and the second corresponds to a correlation or anticorrelation in the direction of emission of the ionized electron and the ionic fragment.

Laboratory-frame asymmetries have been observed in dissociative ionization of D_2 by a CEP-locked-IR laser pulse [13] or by an EUV-IR pump-probe technique [64]. The latter scheme has been used in the theoretical calculations reviewed in the present manuscript. Recently, the role of NWP dynamics in this asymmetry has also been explored for H_2^+ dissociation using phase-stabilized IR pulses [18]. On the other hand, molecular-frame asymmetries were observed in single-photon EUV dissociative ionization of H_2 and D_2 using circularly polarized light and mediated by autoionization of the Q_2 series [26, 65].

The observation of an asymmetry in the photoionization yields in homonuclear diatomic molecules implies the creation of a coherent superposition of states with different parity. For example, an attosecond pulse can excite or ionize the H_2 molecule leaving behind the ion either in the $1s\sigma_g$ or the $2p\sigma_u$ channels. The electronic wave functions ϕ_g and ϕ_u characterizing these levels in the dissociation limit are linear combinations of $1s$ atomic wave functions centered on each nucleus,

$$\phi_{g,u} = \frac{1}{\sqrt{2}}(\phi_{1s,L} \pm \phi_{1s,R}). \tag{1.10}$$

The localized states, denoted left (L) and right (R) states depending on which nucleus carries the remaining electron, are written

$$\phi_{L,R} = \frac{1}{\sqrt{2}}(\phi_g \pm \phi_u). \tag{1.11}$$

Experimentally, electron localization is observed as a laboratory asymmetry in the ejection of ionic fragments with respect to the laser polarization axis. To observe this localization, the coherent superposition must regard ions with the same final kinetic energies and angular momentum of the outgoing electrons. To understand its origin, we will use the following simplified version of the two-electron wave function of the singly ionized molecule produced by, e.g., a combination of a SAP and an IR pulse

$$\begin{aligned} \Psi = c_1 \left[\phi_g(1)\epsilon l_g(2)\right]_g + c_2 \left[\phi_g(1)\epsilon l_u(2)\right]_u \\ + c_3 \left[\phi_u(1)\epsilon l_u(2)\right]_g + c_4 \left[\phi_u(1)\epsilon l_g(2)\right]_u , \end{aligned} \tag{1.12}$$

with the second electron being described by a function of energy ϵ and angular momentum $l_{u,g}$ [64]. Notice that such combination of pulses can lead to two-electron molecular states of both g and u symmetries and that in turn each symmetry can result from the combination of two molecular orbitals of g and u symmetries. For simplicity in the discussion, the wave function in (1.12) has not been antisymmetrized with respect to electrons 1 and 2. The observation of a localization of the remaining electron on the right or left nucleus in the laboratory frame corresponds to the projection of Ψ on the localized states:

$$\Psi_{(L,R),(g,u)} = \phi_{(L,R)}(1)\epsilon l_{g,u}(2). \tag{1.13}$$

The asymmetry can therefore be expressed as

$$N_L - N_R = \sum_{i=g,u}\left[|\langle\Psi_{L,i}|\Psi\rangle|^2 - |\langle\Psi_{R,i}|\Psi\rangle|^2\right] = 4\mathrm{Re}[c_1c_4^* + c_2c_3^*]. \tag{1.14}$$

Note that in terms of the coefficients in (1.12), a laboratory-frame asymmetry is formed by a mixed-parity superposition in which the continuum electron has the same angular momentum, l_u or l_g, in both ionic states, with $c_1, c_4 \neq 0$ and, respectively, $c_2, c_3 \neq 0$ [13]. In contrast, a molecular-frame asymmetry is caused by an interference of contributions with equal parity: the first and third terms in (1.12) for $c_1, c_3 \neq 0$ or the second and fourth terms, for $c_2, c_4 \neq 0$ [26].

1.3.4.2 Laboratory-Frame Asymmetry

An accurate theoretical description of electron localization in H_2 after numerical integration of the TDSE requires the splitting of the proton flux into protons going to the right detector ($H{+}H^+$) and protons going to the left detector ($H^+{+}H$) by projecting onto states of the following form:

$$|R\rangle, |L\rangle = \frac{1}{\sqrt{2}}\sum_{\mathrm{lm}} \mathrm{i}^l \mathrm{e}^{-\mathrm{i}\sigma_l(\epsilon)} Y_{\mathrm{lm}}^*(\mathbf{k})\big[\psi_g^{\epsilon l}(\mathbf{r}_1, \mathbf{r}_2; R)\chi_{\nu_g}(R)\mathrm{e}^{-\mathrm{i}W_g t}$$
$$\pm\psi_u^{\epsilon l}(\mathbf{r}_1, \mathbf{r}_2; R)\chi_{\nu_u}(R)\mathrm{e}^{-\mathrm{i}W_u t}\big], \tag{1.15}$$

where $\sigma_l(\epsilon) = \arg\Gamma(l + 1 - \mathrm{i}/\sqrt{2\epsilon})$ is the Coulomb phase shift, Y_{lm} is a spherical harmonic, $\mathbf{k}$ is the electron momentum, $\{\psi_\alpha^{\epsilon l}(\mathbf{r}_1, \mathbf{r}_2, R)\chi_{\nu_\alpha}(R)\}$ are the eigenstates of the electronic continuum defined in (1.4) and (1.5), and the signs $+$ and $-$ correspond to $|R\rangle$ and $|L\rangle$, respectively [66]. The corresponding fully differential dissociative ionization probability (differential in electron energy, electron emission direction, proton energy, and proton emission hemisphere) is then given by

$$\frac{\mathrm{d}^3 P^d}{\mathrm{d}E_{H^+}\mathrm{d}\epsilon\mathrm{d}\Omega} = \frac{1}{2}\left|\sum_{\alpha,l,m} \mathrm{i}^{-l}\mathrm{e}^{\mathrm{i}\sigma_l(\epsilon)} Y_{\mathrm{lm}}(\Omega) C_{\alpha\nu_\alpha}^{\epsilon l}\right|^2, \tag{1.16}$$

where d stands for either R or L.[1] Here we omit the time dependence of the coefficients, since we take their value at the end of the propagation, where

[1] We set the R direction by $C_{(\alpha=u)\nu_\alpha}^{\epsilon l} \to C_{u\nu_\alpha}^{\epsilon l}$ and the L direction by $C_{(\alpha=u)\nu_\alpha}^{\epsilon l} \to -C_{u\nu_\alpha}^{\epsilon l}$.

$C^{\epsilon l}_{\alpha v_\alpha}(t_f) \sim C^{\epsilon l}_{\alpha v_\alpha}(\infty)$. Notice that, in writing (1.15), we have taken into account that for photon energies below ∼35 eV, the R and L paths mainly result from the combination of the $1s\sigma_g$ and $2p\sigma_u$ H_2^+ molecular states into 1s orbitals localized in just one of the protons (see (1.11)).

When one wants to compare with experiments in which the energy of the electron is not measured, the fully differential dissociative ionization probabilities (1.16) must be integrated over the electron kinetic energy

$$\frac{d^2 P^d}{dE_{H^+} d\Omega} = \frac{1}{2} \int d\epsilon \left| \sum_{\alpha,l,m} i^{-l} e^{i\sigma_l(\epsilon)} Y_{lm}(\Omega) C^{\epsilon l}_{\alpha v_\alpha} \right|^2 . \tag{1.17}$$

An additional integration over the solid angle gives the single differential probabilities in proton (or deuterium) kinetic energy

$$\frac{dP^d}{dE_{H^+}} = \frac{1}{2} \int d\epsilon \sum_l \sum_{\alpha,\alpha'} C^{\epsilon l *}_{\alpha' v'_\alpha} C^{\epsilon l}_{\alpha v_\alpha} . \tag{1.18}$$

which, for a given direction of the ionic fragment, can be rewritten as

$$\frac{dP^{R,L}}{dE_{H^+}} = \frac{1}{2} \int d\epsilon \sum_l \left(|C^{\epsilon l}_{g v_g}|^2 \pm 2\mathrm{Re}[C^{\epsilon l *}_{g v_g} C^{\epsilon l}_{u v_u}] + |C^{\epsilon l}_{u v_u}|^2 \right) . \tag{1.19}$$

being the total proton flux as a function of the proton energy

$$\frac{dP}{dE_{H^+}} = \frac{dP^R}{dE_{H^+}} + \frac{dP^L}{dE_{H^+}} = \int d\epsilon \left(|C^{\epsilon l}_{g v_g}|^2 + |C^{\epsilon l}_{u v_u}|^2 \right) . \tag{1.20}$$

In order to compare with existing experiments, one has to evaluate the proton asymmetry parameter defined as

$$\beta(E_{H^+}) = \frac{\left(\frac{dP^L}{dE_{H^+}} - \frac{dP^R}{dE_{H^+}} \right)}{\left(\frac{dP^L}{dE_{H^+}} + \frac{dP^R}{dE_{H^+}} \right)} . \tag{1.21}$$

Using (1.19) and (1.20), we obtain

$$\beta(E_{H^+}) = 2 \int d\epsilon \sum_l \left(\mathrm{Re}[C^{\epsilon l *}_{g v_g} C^{\epsilon l}_{u v_u}] \right) / \frac{dP}{dE_{H^+}} . \tag{1.22}$$

Note that the asymmetry arises from the interference between the $1s\sigma_g$ and $2p\sigma_u$, as is explicitly seen in the numerator and that, apart from the renormalization factor that appears in the denominator, (1.14) is a particular case of the above equation.

Regarding the outgoing electron flux, (1.17) gives the Molecular-frame photoelectron angular distributions (MFPADs) for the $H+H^+$ and $H^+ + H$ processes, so we may also define an electron asymmetry parameter as

$$\alpha^d(E_{H^+}) = \frac{\left(\frac{\mathrm{d}P^{\mathrm{d}l}}{\mathrm{d}E_{H^+}} - \frac{\mathrm{d}P^{\mathrm{d}r}}{\mathrm{d}E_{H^+}}\right)}{\left(\frac{\mathrm{d}P^{\mathrm{d}l}}{\mathrm{d}E_{H^+}} + \frac{\mathrm{d}P^{\mathrm{d}r}}{\mathrm{d}E_{H^+}}\right)}, \tag{1.23}$$

where

$$\frac{\mathrm{d}P^{\mathrm{d}r}}{\mathrm{d}E_{H^+}} = \int_0^{\pi/2} \sin\theta \, \mathrm{d}\theta \int_0^{2\pi} \mathrm{d}\phi \frac{\mathrm{d}^2 P^d}{\mathrm{d}E_{H^+}\mathrm{d}\Omega}, \tag{1.24}$$

$$\frac{\mathrm{d}P^{\mathrm{d}l}}{\mathrm{d}E_{H^+}} = \int_{\pi/2}^{\pi} \sin\theta \, \mathrm{d}\theta \int_0^{2\pi} \mathrm{d}\phi \frac{\mathrm{d}^2 P^d}{\mathrm{d}E_{H^+}\mathrm{d}\Omega}. \tag{1.25}$$

Equation (1.23) defines the asymmetry related to the electron angular distributions for protons escaping to the left α^L or to the right α^R. Of course, this asymmetry depends on the proton kinetic energy and also on the time delay between the pump and probe pulses, i.e., $\alpha^d = \alpha^d(E_{H^+}, \tau)$.

1.3.4.3 Photon Energy Regimes for Photoionization of H_2 and D_2 Molecules

A brief analysis of the potential energy curves for H_2 (and D_2) molecules shown in Fig. 1.3 allows one to predict which are the single ionization channels that are open when a given amount of energy is absorbed. As electronic excitations upon EUV/XUV photon absorption are practically instantaneous compared to the nuclear motion (*vertical* transitions), the most probable transitions will be those with a high spatial overlap between the initial and final vibrational wave functions. The spatial region accessible from the $\nu = 0$ ground state of the molecule is labeled as the Franck–Condon region in Fig. 1.3. In single photoionization of H_2 (D_2), the dominant channel corresponds to leaving the molecular ion in its bound vibrational states. However, depending on the photon energy and the polarization of the light with respect to the molecular axis, other channels can also be observed. For low-energy photons, ($E_\gamma < 25\,\mathrm{eV}$), dissociative ionization associated to the $1s\sigma_g$ state releases a small fraction (2% of the total ionization yield) of low-kinetic-energy ionic fragments ($E_\mathrm{k} < 1\,\mathrm{eV}$, see Fig. 1.3) due to the rapid decrease of the Franck–Condon overlap between the initial vibrational state and the final dissociative state. For photon energies above 25 eV, autoionization through different

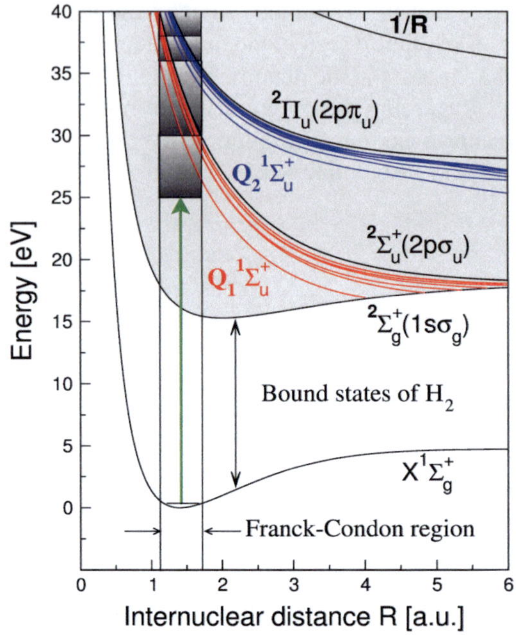

Fig. 1.3 Potential energy curves of the H_2 molecule. The *shadowed areas* show the thresholds of the different regimes for ionization (see text)

doubly excited states can also occur, contributing up to 5% to the total ionization probability. Autoionization is clearly distinguishable from direct ionization in the dissociative ionization spectra because it always leads to protons with $E_k > 1$ eV. The contribution of different doubly excited states varies with (1) the symmetry of the final states, which depends on the light polarization (linear, circular) with respect to the molecular axis (parallel, perpendicular or combinations of them); (2) the oscillator strength associated to their electronic dipole couplings with the ground state; and (3) the available energy.

For instance, for photon energies of 25–36 eV and linearly polarized light parallel to the molecular axis, only the $Q_1\,^1\Sigma_u^+$ series of doubly excited states is effectively populated. Autoionization through these states produces ionic fragments with energies between 2 and 7 eV [67, 68]. However, one-photon absorption above 31 eV with linearly polarized light aligned perpendicular to the molecule populates the $Q_2\,^1\Pi_u$ doubly excited states which autoionize into both the $1s\sigma_g$ and $2p\sigma_u$ states, resulting in ionic fragments with kinetic energies of 1–5 eV and 5–8 eV, respectively [61]. Direct photoionization to the repulsive $2p\sigma_u$ electronic state of the ion becomes accessible for photon energies >30 eV. Finally, for relatively high photon energies, >38 eV, higher dissociative electronic states of the ion are reached, leading to high-energy fragments ($E_k = 5-10$ eV). Experimental dissociative ionization spectra are sometimes obtained for fragments released in a cone around the laser polarization axis, so that the involvement of doubly excited states of different symmetries cannot be ruled out a priori.

1.4 H_2 and D_2 Dissociative Ionization by Two-Color (XUV+IR) Fields

In this section, we will focus on results for dissociative ionization of hydrogenic molecules through XUV pump–IR probe schemes. In the first part, we use a XUV SAP in combination with an IR pulse. Ionization yields and fully differential electron and proton distributions are explored for different time delays. In the second part, we study the influence of using an APT instead of a single pulse.

1.4.1 Single Attosecond Pulse plus Infrared: Electron Localization

Electron localization in the D_2 molecule can be controlled by inducing ionization with a SAP and driving the ionized electron in an IR field. The experimental and theoretical setups have been described in [64]: the SAP has a central frequency of 30 eV, an effective pulse length of 400 as, and a peak intensity of 10^9 W/cm^2. The IR pulse is a 6-fs FWHM 750-nm field with a peak intensity of $3 \cdot 10^{12}$ W/cm^2. Both fields are collinearly polarized with the internuclear axis. As shown in Fig. 1.3, a 30-eV photon absorption ionizes the target leaving the ion mainly in its $1s\sigma_g$ electronic ground state and populates the doubly excited states of the $Q_1 \Sigma_u^+$ series. The broad bandwidth of the attosecond pulse populates a wide energy region, thus also reaching the second ionization threshold $2p\sigma_u$, although with a smaller probability. The presence of the IR field modifies the ionization probability ratios, which will depend on the time delay between the EUV and the IR pulses, and also allows for ionization through states of Σ_g^+ symmetry, which are optically forbidden in single-photon absorption.

The calculated proton kinetic energy distributions as a function of the relative delay between the pulses are shown in Fig. 1.4 for H_2 photoionization. The IR pulse affects the kinetic energy distributions of the ion fragments: for low-energy photons ($E_k < 1$ eV), the IR causes bond-softening (BS) of the bound $1s\sigma_g$ vibrational wave packet, enhancing the ionization probability. This effect peaks at $\tau = +10$ fs, when the wave packet is near the outer turning point of the potential energy curve [69,70]. Above 1 eV and for long time delays ($\tau > 8$ fs), the kinetic energy distribution is almost independent of the delay.

When both pulses overlap ($\tau \sim 0$ fs), the ion signal increases strongly for kinetic energies of $\sim$ 8 eV (white rectangle) and decreases at intermediate energies (3 eV $< E_k <$ 5 eV). This enhancement is due to the IR-induced mixing of the $1s\sigma_g$ and $2p\sigma_u$ states: electrons in the $1s\sigma_g$ channel can absorb IR photons to populate the $2p\sigma_u$ channel. The opposite transition, $2p\sigma_u \rightarrow 1s\sigma_g$, is also possible, but due to the much smaller probability of $2p\sigma_u$, there is a net increase in the $2p\sigma_u$ probability from the first process. Photoionization through the $Q_1{}^1\Sigma_u^+$ autoionizing states also

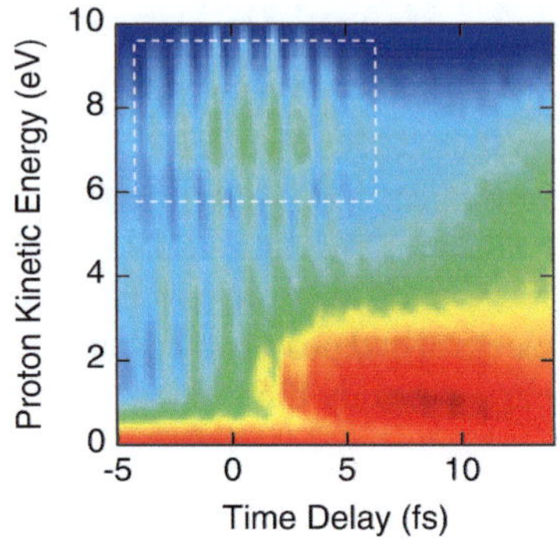

Fig. 1.4 Proton kinetic energy distribution as a function of the delay between the attosecond pulse and the IR pulse, for the SAP+IR scheme. The *color scale* shows fragment yield in arbitrary units. The white rectangle points to the region in which autoionization is visible

contributes to this enhancement, for relative time delays shorter than their expected averaged lifetimes ($\simeq$ 3 fs [67]).

Besides the observed enhancement for specific proton kinetic energies, an oscillating pattern appears for the KER probabilities as a function of the time delay. This oscillation can be directly related to electron localization and, consequently, it can be analyzed in terms of asymmetries in the dissociative ionization process, as explained in Sect. 1.3.4. The calculated asymmetry parameter for the formation of H^+ fragments, (1.22), is shown in Fig. 1.5 as a function of the proton kinetic energy and time delay. An oscillation can be observed along almost the entire kinetic energy range in which H^+ ions are formed, with a periodicity of T, the IR period. The phase of the oscillations strongly depends on the kinetic energy of the fragment that is measured.

We saw in Sect. 1.3 that the observed asymmetry can be easily understood in terms of the simplified two-electron wave function of singly ionized H_2 (1.12). The laboratory-frame asymmetry appears when there is a superposition between two states with different ionic states, $1s\sigma_g$ and $2p\sigma_u$, but the same parity in the angular momentum of the ejected electron, either l_u or l_g. The first ionic state is even with respect to exchange of the nuclei, while the second is odd, so any superposition leads to electron localization in one of the nuclei and therefore to an asymmetry in the ion detection.

The theoretical results for the asymmetry parameter plotted in Fig. 1.5 are in very good agreement with the experimental measurements reported in [64]. Electron localization is very pronounced for delays of up to 7 fs and kinetic energies above 5 eV, but it is hardly visible for longer delays and lower kinetic energies. This can be explained in terms of the coefficients in (1.12). The asymmetry was given in (1.14) by

$$N_L - N_R = 4\mathrm{Re}[c_1 c_4^* + c_2 c_3^*]. \tag{1.26}$$

In the absence of the infrared pulse, the EUV one-photon ionization from the ground state of H_2 produces a final molecular state with odd symmetry, Σ_u^+; therefore, only c_2 and c_4 are nonzero, and no laboratory-frame asymmetry is observed. However, the IR field induces a change on the total wave function, populating Σ_g^+ states.

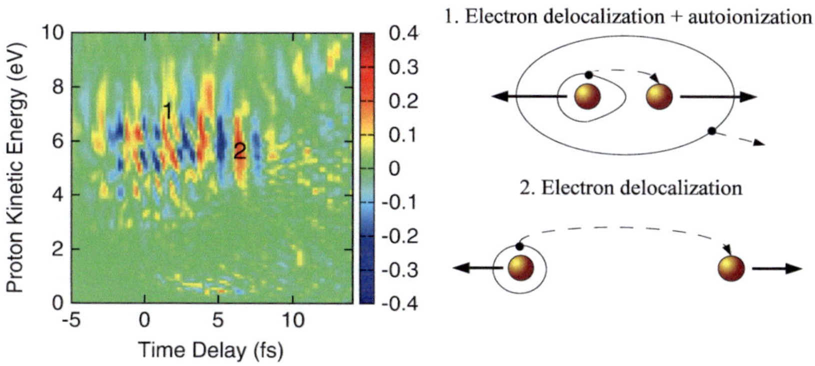

Fig. 1.5 Asymmetry in EUV-IR dissociative ionization of hydrogen [64]: calculated asymmetry parameter (*color scale*) for the formation of H^+ ions, as a function of the ion kinetic energy E_k and the pump-probe delay. *Right*: mechanisms that lead to asymmetry in VUV-IR dissociative ionization: the regions 1 and 2 in the kinetic energy spectrum relate to the mechanisms described in the right

This modifies the electronic continuum involving doubly excited states (mechanism 1 in Fig. 1.5) and also affects the molecular ion (mechanism 2). The effect of the IR pulse right after the photoexcitation process, but before autoionization takes place, leads to mechanism 1, in which the remaining electron oscillates between the two protons following the IR field in the presence of the second electron that has not yet been ionized (as mentioned above, autoionization does not occur after a few femtoseconds). For longer time delays (mechanism 2), the IR pulse leads to similar oscillations of the remaining electron during the dissociation process, but this oscillations are no longer perturbed by the other electron, which is already gone. This is the reason why oscillations in the asymmetry parameter are smoother in this region of the spectrum.

The shape of the fringes arising from mechanism 2 can be understood by means of the semiclassical Landau–Zener model, which describes the IR-induced population transfer between the $1s\sigma_g \equiv \varphi_1$ and $2p\sigma_u \equiv \varphi_2$ levels in terms of the so-called quasistatic states, which are eigenstates of the two-level problem in the presence of a (static) electric field,

$$\begin{aligned}\Psi_1 &= \cos[\theta(t)]\varphi_1 + \sin[\theta(t)]\varphi_2, \\ \Psi_2 &= -\sin[\theta(t)]\varphi_1 + \cos[\theta(t)]\varphi_2.\end{aligned} \tag{1.27}$$

The transition frequency $\theta(t)$ is related to the energy splitting $\omega_0(R)$ between the two states and to the IR-induced dipole coupling $V_{g,u}(R,t) = -\mu(R)E(t)$, where $\mu(R)$ is the electronic dipole moment, via

$$\tan[2\theta(t)] = -2V_{g,u}(R,t)/\omega_0(R). \tag{1.28}$$

During photoexcitation, $\omega_0(R) \gg |V_{g,u}(R,t)|$ and the transition probability is small, so the NWP remains in Ψ_2. As dissociation takes place and R increases, $\omega_0(R)$ becomes comparable to the dipole coupling, and a coherent superposition of the quasistatic states Ψ_L and Ψ_R is formed. Near the end of the dissociation, the internuclear potential barrier becomes so high that the electron cannot switch from one nucleus to the other and localization is fixed; in other words, here $\omega_0(R) \ll |V_{g,u}(R,t)|$ and the NWP switches between the two quasistatic states. Since this localization process depends on the internuclear distance at which the NWP is launched, the asymmetry has a correlated dependence on E_k and τ [64].

1.4.2 Attosecond Pulse Train plus Infrared: Role of Different States

The use of an APT instead of a single pulse in combination with the IR field substantially changes the physics. The APT spectrum has peaks at several harmonics of the generating frequency, with a narrower energy bandwidth of ~1 eV (see Figs. 1.1 and 1.6a). The APT contains XUV photons which correspond to odd harmonics of the IR frequency, but absorption or emission of photons from the IR field can lead to final energies corresponding to even multiples of the IR frequency. The existence of different paths reaching the same final state, Fig. 1.6b, leads to interference effects in the ionization distributions, see Sect. 1.2.

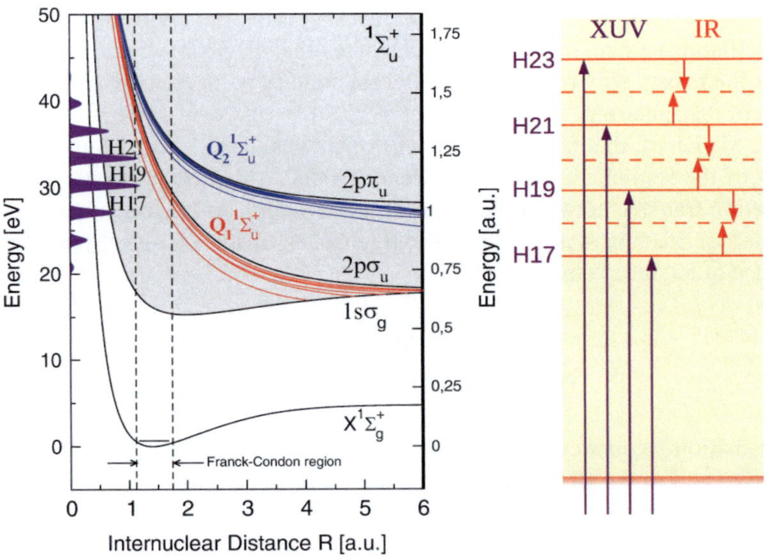

Fig. 1.6 *Left*: scheme of the APT used in the calculations. *Right*: absorption of XUV photons from different XUV harmonics followed by IR absorption or emission leads to interferences

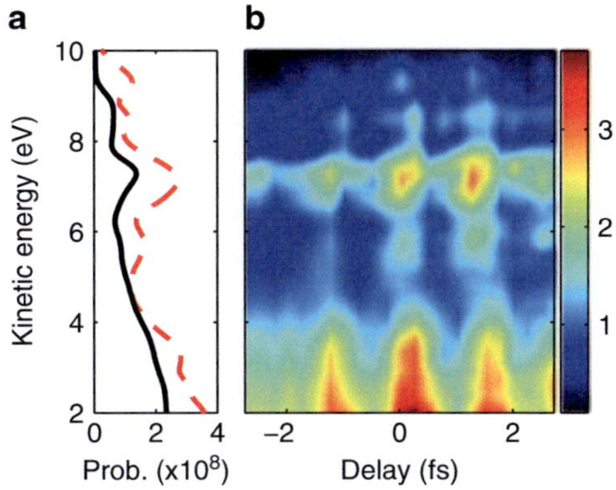

Fig. 1.7 (**a**) Fragment kinetic energy spectra from dissociative ionization of D_2 by the APT only (*black solid lines*) and by the APT+IR (cycle averaged) at temporal overlap (*red dashed lines*) for linear polarization. (**b**) Time dependence of the KER as a function of delay τ between the APT and the IR pulse

The ATP+IR scheme hereby discussed for D_2 ionization involves a 4-pulse APT and a 2.8-fs FWHM 780-nm field with a peak intensity of 3×10^{12} W/cm^2 [71]. The energetics of the problem is schematically represented in Fig. 1.6. The APT spectrum is chosen so that it allows direct ionization through the first excited ionic channel $2\text{p}\sigma_u$, while it limits ionization through channels associated with higher ionic states as much as possible. We will focus here on results for parallel polarization, although similar results for perpendicular polarization can be found in [71].

The D^+ kinetic energy spectrum resulting from the interaction of D_2 with the APT alone is shown in black in Fig. 1.7a. Several contributions can be distinguished: direct ionization to the $1\text{s}\sigma_g$ state produces stable D_2^+ and a small contribution from vibrational dissociative states (2%) that originates low kinetic energy deuterons ($E_\text{k} < 1$ eV). The main contribution between 1 and 5 eV comes from autoionization of the $^1\Sigma_u^+ Q_1$ doubly excited states [61] (red curves in Fig. 1.6a). Fragments with higher kinetic energies (7–8 eV) come from multiple dynamical pathways of which direct ionization to the $2\text{p}\sigma_u$ continuum dominates for parallel polarization.

The addition of an IR field causes significant changes in the spectrum. This is shown in Fig. 1.7a with a red dashed line, averaged for the XUV-IR time delay over one full IR cycle. For low kinetic energies, the enhancement is due to bond softening, and there is an enhancement for higher kinetic energies ($\sim$7 eV), which corresponds to an increase in the $2\text{p}\sigma_u$ probability. These two features were already present in the SAP+IR case, Sect. 1.4.1.

The dependence of the D^+ kinetic energy distributions with the APT-IR delay is shown in Fig. 1.7b. The ionization probability oscillates with the time delay for all kinetic energies with a periodicity of $T/2$, where T is the IR period. These results are reminiscent of the work by Johnsson et al. [52] in helium ionization, where similar oscillations were found in the ionization probability as a function of the APT-IR delay that were attributed to interference between electron wave packets [32, 55]. In the case of D_2 shown here, however, the origin of the oscillations is different, since they appear both for APT+IR and for SAP+IR. Moreover, at

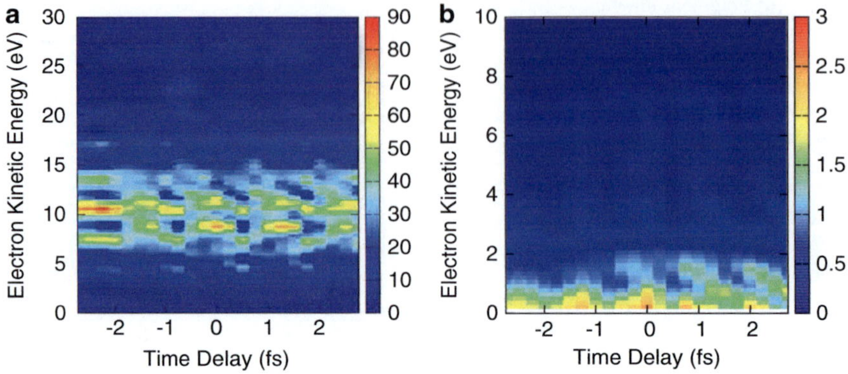

Fig. 1.8 (**a**) Electron kinetic energy (EKE) spectra for D_2 ionization as a function of the time delay between the APT and the IR pulse. (**b**) Partial contribution, to the EKE spectra shown on (**a**), from electronic continuum states associated to only the $2p\sigma_u$ ionization threshold. All probabilities are multiplied by 10^8

variance with [52], all the harmonics associated with the APT used in this work lie in the electronic continuum.

For a complete picture of the ionization channels, we show in Fig. 1.8a the electron kinetic energy distributions (therefore, sum/integrated over all vibrational states) as a function of the time delay. For the most negative time delays ($\tau < -2$ fs), the electron distributions exhibit peaks that result from absorption of the different harmonics contained in the APT. Since our theoretical treatment allows for a systematic analysis of the contributions from each ionization thresholds and symmetries of the final states to the ionization probabilities, we also show on Fig. 1.8b the ionization probability contribution from the continuum associated to the $2p\sigma_u$. From the comparison, it can be seen that the total ionization yield mainly comes from the continuum states associated to the $1s\sigma_g$ state of the ion. But it should be remarked that, as demonstrated by a simple analysis using the Franck–Condon principle, the $1s\sigma_g$ contribution leads to low proton energies ($E_k < 1$ eV) or bound vibrational states of the ion, whereas ionization through the $2p\sigma_u$ ionic state leads to protons of significantly higher energies. Thus, the peak appearing at high proton energies (7 eV) in Fig. 1.7 comes from final states associated to the $2p\sigma_u$.

According to the dipole selection rules, the final continuum state can have a total symmetry (ion+electron) Σ_g or Σ_u depending on the number of absorbed photons and following the pathway

$$\psi_0(\Sigma_g) \overset{\omega}{\longleftrightarrow} \Sigma_u \overset{\omega}{\longleftrightarrow} \Sigma_g. \tag{1.29}$$

Thus, the complicated patterns observed in the electron distributions shown in Fig. 1.8a for $\tau > -2$ fs can be unraveled by plotting the separate contribution of final states with Σ_u and Σ_g symmetries. This is shown in Fig. 1.9. One-photon absorption from the ground state can only lead to Σ_u final states, which explains the

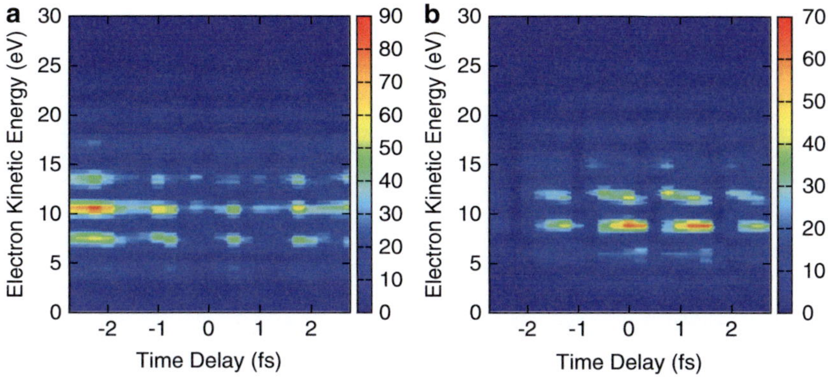

Fig. 1.9 Electron kinetic energy spectra for D_2 ionization as a function of time delay between the APT and the IR pulse, for total final symmetry Σ_u^+ (**a**) and Σ_g^+ (**b**). All probabilities are multiplied by 10^8

distributions for negative time delays: in the absence of IR (or hardly overlapping with the APT), the Σ_u states leave a more intense signal revealing the harmonics of the APT. However, when APT and IR fields overlap, and for positive time delays, the Σ_g final states are significantly populated at electron energies corresponding to the sidebands of the APT harmonics (i.e., through absorption of an even number of photons of the combined APT+IR fields). The sidebands are located in between the peaks associated to the odd harmonics of the APT, and their intensity oscillates with the APT-IR delay with half the periodicity of the IR field, $T/2$. This is similar to what has been observed in RABBITT experiments [37] carried out in atoms.

We now analyze the high-energy deuterons ($E_\mathrm{k} \simeq 7\,\mathrm{eV}$) that correspond to the $2\mathrm{p}\sigma_u$ ionization continuum. Figure 1.10 shows the contributions of the Σ_g and Σ_u final states to the $2\mathrm{p}\sigma_u$ channel (the sum of them was shown in Fig. 1.8b). The largest contribution corresponds to the Σ_u final symmetry, Fig. 1.10a, originated by direct absorption of the higher harmonics of the APT pulse that are energetic enough to reach the $2\mathrm{p}\sigma_u$ state by a direct transition, see Fig. 1.6a.

To understand the origin of the oscillations observed at higher D^+ kinetic energies, we use a FNA for solving the TDSE, with the same field as in the full calculation. The total wave function is now expanded in a basis of electronic states that only includes the D_2 ground state and the wave functions of the continuum associated to the ground (gerade) and first excited (ungerade) states of the molecular ion,

$$\Psi(t) = c_0(t)\psi_0 + \sum_l \int \mathrm{d}\epsilon\, c_{g,\epsilon l}(t)\psi_{g,\epsilon l} + \sum_l \int \mathrm{d}\epsilon\, c_{u,\epsilon l}(t)\psi_{u,\epsilon l}, \tag{1.30}$$

where l is the index for the partial wave of the electron in the continuum of energy ϵ. The internuclear distance is set at $R = 1.7$ a.u., instead of at the equilibrium distance

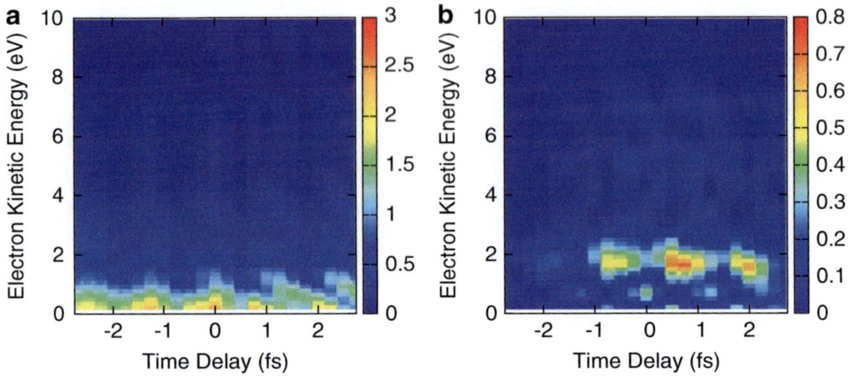

Fig. 1.10 Contributions to the EKE spectra associated to the 2pσ_u ionization threshold, for total symmetry Σ_u^+ (**a**) and Σ_g^+ (**b**). The sum of both contributions is shown in Fig. 1.8b. All probabilities are multiplied by 10^8

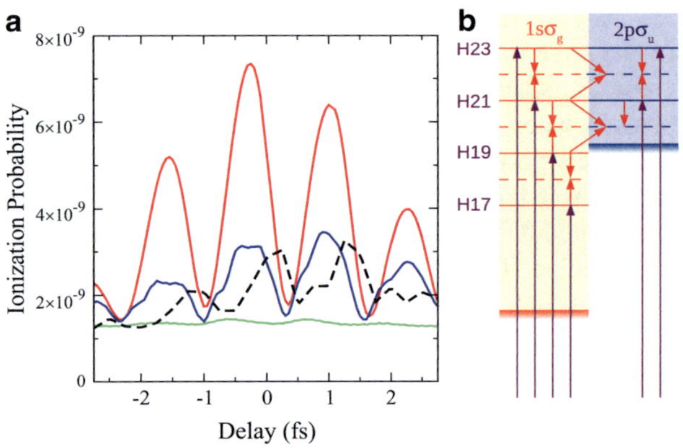

Fig. 1.11 (**a**) Probability of ionization to the 2pσ_u continuum at an IR intensity of $3 \cdot 10^{12}$ W/cm^2. The black dashed line is the result of the full calculation including nuclear motion. The other curves are FNA model with all couplings turned on (*blue line*), only couplings within the $2p\sigma_u$ continuum (*green line*), and only couplings between the $1s\sigma_g$ and $2p\sigma_u$ continua (*red line*). (**b**) Schematic illustration of APT+IR ionization to the $1s\sigma_g$ (*orange*) and $2p\sigma_u$ continuum (*blue*). Purple arrows represent ionization by the odd harmonics of the APT, red vertical arrows IR-induced couplings within the continua, and red diagonal arrows the coupling between the continua

of the neutral, 1.4 a.u., so that the H19 harmonic in the APT (the most intense one) lies above the 2pσ_u state, which is the channel associated to the high-energy protons.

In Fig. 1.11, results from the FNA calculations (using three different truncations in the basis of electronic states) are compared with those obtained in the full calculation. In the figure, the total ionization probability that remains in the continuum associated to the 2pσ_u ionic state is plotted as a function of the APT-IR time delay; the black dashed line represents the results from the full-dimensional calculation. The first FNA calculation (green line) only includes the electronic continuum associated to the $2p\sigma_u$ state of the ion. The sidebands of the APT harmonics appear in its photoelectron spectrum, but no oscillation is found in the total ionization yield. The second FNA calculation (red line) includes the coupling between the $1s\sigma_g$ and $2p\sigma_u$ continua (purple and red diagonal lines in Fig. 1.11b) but without the couplings within the individual continua: in this case, the odd harmonics primarily access the $1s\sigma_g$ continuum, and the sidebands in the $2p\sigma_u$ continuum can be accessed from those states through absorption or emission of IR photons. This originates constructive or destructive interferences which lead to the oscillations in the total yield associated to the $2p\sigma_u$ ionic state, with a period of half an optical IR cycle. Finally, the FNA calculation including both continua and all couplings (blue line) leads to a result that closely resembles that of the full calculation, albeit there is a phase difference, which can be attributed to the neglection of nuclear motion in the FNA model.

1.5 Conclusions and Outlook

In this chapter, we have described a method to study electron and nuclear dynamics of diatomic (H_2 and D_2) molecules interacting with ultrashort laser pulses. This is done with a close-coupling method which includes all electronic and vibrational (dissociative) degrees of freedom, as well as the effect of electron correlation and interferences between different ionization and dissociation pathways. In particular, we have analyzed electron localization in H_2 dissociation through a particular observable, the laboratory-frame asymmetry, which relates the fragment yield onto opposite directions in the laboratory frame. For this purpose, we have considered the following scheme: a SAP ionizes the H_2 molecule, and the resulting H_2^+ molecule in the intense IR field is in a superposition of gerade and ungerade states, which eventually leads to electron localization [64]. The asymmetry coefficient as a function of the SAP-IR delay shows an oscillation that can be attributed to two different mechanisms. The first is due to interferences between two states populated at the photoexcitation process (the $2p\sigma_u$ dissociating state, and the Q_1 autoionizing states, which decay to the $1s\sigma_g$ state). The second mechanism is due to the influence of the IR field at the dissociation channel, where it couples the $2p\sigma_u$ dissociating channel with the $1s\sigma_g$ channel.

In a second scheme, we have replaced the SAP by an APT, which strongly modifies the initial spectral distribution and leads to new interferences in the ionization continua. In this APT+IR setup, the ionization yields show oscillatory patterns that are due to the interference between channels coming from absorption of an XUV photon followed from absorption or emission of an IR photon. These interferences also lead to RABITT-type sidebands, which are obscured in the spectra of the nuclear fragments but are clearly visible in the electron kinetic energy distributions.

In both schemes, the ab initio theoretical treatment presented in this chapter is necessary to understand the ionization dynamics of H_2 and D_2 induced by ultrashort pulses and observed in recent experiments. In particular, it opens the way to unravel the contributions from different ionization channels and to provide a time-resolved picture of this dynamics. And, as it has been shown, this is only possible by means of a time-dependent treatment that describes the combined motion of electrons and nuclei.

References

1. E. Goulielmakis, M. Schultze, M. Hofstetter, V. Yakovlev, J. Gagnon, M. Uiberacker, A.L. Aquila, E.M. Gullikson, D.T. Attwood, R. Kienberger, F. Krausz, U. Kleineberg, Science **320**, 1614 (2008)
2. H. Mashiko, A. Suda, K. Midorikawa, Opt. Lett. **29**, 1927 (2004)
3. A. Barty, J. Phys. B Atom. Mol. Opt. Phys. **43**, 194014 (2010)
4. C. Bostedt, H.N. Chapman, J.T. Costello, J.R. Crespo López-Urrutia, S. Düsterer, S.W. Epp, J. Feldhaus, A. Föhlisch, M. Meyer, T. Möller, Nuclear Instruments and Methods in Physics Research Section A: Accelerators Spectrometers Detectors Associated Equipment **601**, 108 (2009)
5. M. Uiberacker, T. Uphues, M. Schultze, A.J. Verhoef, V. Yakovlev, M.F. Kling, J. Rauschenberger, N.M. Kabachnik, H. Schröder, M. Lezius, K.L. Kompa, H.G. Muller, M.J.J. Vrakking, S. Hendel, U. Kleineberg, U. Heinzmann, M. Drescher, F. Krausz, Nature **446**, 627 (2007)
6. M. Drescher, M. Hentschel, R. Kienberger, M. Uiberacker, V. Yakovlev, A. Scrinzi, T. Westerwalbesloh, U. Kleineberg, U. Heinzmann, F. Krausz, Nature **419**, 803 (2002)
7. J. Itatani, J. Levesque, D. Zeidler, H. Niikura, H. Pépin, J.C. Kieffer, P.B. Corkum, D.M. Villeneuve, Nature **432**, 867 (2004)
8. G. Yudin, S. Chelkowski, J. Itatani, A.D. Bandrauk, P.B. Corkum, Phys. Rev. A **72**, 051401(R) (2005)
9. S. Chelkowski, G. Yudin, A.D. Bandrauk, J. Phys. B Atom. Mol. Opt. Phys. **39**, S409 (2006)
10. M.H. Xu, L.Y. Peng, Z. Zhang, Q. Gong, J. Phys. B Atom. Mol. Opt. Phys. **44**, 021001 (2011)
11. H. Niikura, F. Légaré, R. Hasbani, A.D. Bandrauk, M.Y. Ivanov, D.M. Villeneuve, P.B. Corkum, Nature **417**, 917 (2002)
12. H. Niikura, F. Légaré, R. Hasbani, M.Y. Ivanov, D.M. Villeneuve, P.B. Corkum, Nature **421**, 826 (2003)
13. M.F. Kling, C. Siedschlag, A.J. Verhoef, J.I. Khan, M. Schultze, T. Uphues, Y. Ni, M. Uiberacker, M. Drescher, F. Krausz, M.J.J. Vrakking, Science **312**, 246 (2006)
14. X.M. Tong, C.D. Lin, Phys. Rev. Lett. **98**, 123002 (2007)
15. A. Staudte, D. Pavičić, S. Chelkowski, D. Zeidler, M. Meckel, H. Niikura, M. Schöffler, S. Schössler, B. Ulrich, P. Rajeev, T. Weber, T. Jahnke, D.M. Villeneuve, A.D. Bandrauk, C.L. Cocke, P.B. Corkum, R. Dörner, Phys. Rev. Lett. **98**, 073003 (2007)
16. S. Baker, J.S. Robinson, C.A. Haworth, H. Teng, R.A. Smith, C. ChirilÄ, M. Lein, J.W.G. Tisch, J.P. Marangos, Science **312**, 424 (2006)
17. M. Magrakvelidze, F. He, T. Niederhausen, I.V. Litvinyuk, U. Thumm, Phys. Rev. A **79**, 033410 (2009)
18. B. Fischer, M. Kremer, T. Pfeifer, B. Feuerstein, V. Sharma, U. Thumm, C.D. Schröter, R. Moshammer, J. Ullrich, Phys. Rev. Lett. **105**, 223001 (2010)
19. A. Giusti-Suzor, F.H. Mies, L.F. Dimauro, E. Charron, B. Yang, J. Phys. B Atom. Mol. Opt. Phys. **28**, 309 (1995)
20. A.D. Bandrauk, S. Chelkowski, Phys. Rev. Lett. **87**, 273004 (2001)
21. A. Palacios, H. Bachau, F. Martín, J. Phys. B Atom. Mol. Opt. Phys. **38**, L99 (2005)
22. A. Palacios, S. Barmaki, H. Bachau, F. Martín, Phys. Rev. A **71**, 063405 (2005)
23. F. He, C. Ruiz, A. Becker, Phys. Rev. Lett. **99**, 083002 (2007)

24. C. Calvert, W. Bryan, W. Newell, I.D. Williams, Phys. Rep. **491**, 1 (2010)
25. S. Gräfe, M.Y. Ivanov, Phys. Rev. Lett. **99**, 163603 (2007)
26. F. Martín, J. Fernández, T. Havermeier, L. Foucar, T. Weber, K. Kreidi, M. Schöffler, L. Schmidt, T. Jahnke, O. Jagutzki, A. Czasch, E.P. Benis, T. Osipov, A.L. Landers, A. Belkacem, M.H. Prior, H. Schmidt-Böcking, C.L. Cocke, R. Dörner, Science **315**, 629 (2007)
27. A. Scrinzi, M.Y. Ivanov, R. Kienberger, D.M. Villeneuve, J. Phys. B Atom. Mol. Opt. Phys. **39**, R1 (2006)
28. R. Kienberger, M. Uiberacker, M.F. Kling, F. Krausz, J. Mod. Opt. **54**, 1985 (2007)
29. M.F. Kling, M.J.J. Vrakking, Ann. Rev. Phys. Chem. **59**, 463 (2008)
30. T. Pfeifer, M.J. Abel, P.M. Nagel, A. Jullien, Z.H. Loh, M. Justine Bell, D.M. Neumark, S.R. Leone, Chem. Phys. Lett. **463**, 11 (2008)
31. F. Krausz, M.Y. Ivanov, Rev. Mod. Phys. **81**, 163 (2009)
32. X.M. Tong, P. Ranitovic, C.L. Cocke, N. Toshima, Phys. Rev. A **81**, 021404(R) (2010)
33. L.V. Keldysh, Sov. Phys. JETP **20**, 1037 (1965)
34. F. He, A. Becker, U. Thumm, Phys. Rev. Lett. **101**, 213002 (2008)
35. N. Takemoto, A. Becker, Phys. Rev. Lett. **105**, 203004 (2010)
36. J. Mauritsson, P. Johnsson, E. Gustafsson, A. LHuillier, K.J. Schafer, M.B. Gaarde, Phys. Rev. Lett. **97**, 013001 (2006)
37. P.M. Paul, E.S. Toma, P. Breger, G. Mullot, F. Auge, P. Balcou, H.G. Muller, P. Agostini, Science **292**, 1689 (2001)
38. S. Aseyev, Y. Ni, L. Frasinski, H. Muller, M. Vrakking, Phys. Rev. Lett. **91**, 223902 (2003)
39. E. Foumouo, P. Antoine, H. Bachau, B. Piraux, New J. Phys. **10**, 025017 (2008)
40. K. Ishikawa, Y. Kawazura, K. Ueda, J. Mod. Opt. **57**, 999 (2009)
41. M. Hentschel, R. Kienberger, C. Spielmann, G.A. Reider, N. Milosevic, T. Brabec, P.B. Corkum, U. Heinzmann, M. Drescher, F. Krausz, Nature **414**, 509 (2001)
42. F.H.M. Faisal, J. Phys. B Atom. Mol. Opt. Phys. **6**, L89 (1973)
43. M. Lewenstein, P. Balcou, M.Y. Ivanov, A. L'Huillier, P.B. Corkum, Phys. Rev. A **49**, 2117 (1994)
44. J. Itatani, F. Quéré, G. Yudin, M.Y. Ivanov, F. Krausz, P.B. Corkum, Phys. Rev. Lett. **88**, 173903 (2002)
45. V. Yakovlev, F. Bammer, A. Scrinzi, J. Mod. Opt. **52**, 395 (2005)
46. A. Palacios, H. Bachau, F. Martín, Phys. Rev. A **74**, 031402(R) (2006)
47. J.F. Pérez-Torres, J.L. Sanz-Vicario, H. Bachau, F. Martín, J. Phys. B Atom. Mol. Opt. Phys. **43**, 015204 (2010)
48. S. Gräfe, V. Engel, M.Y. Ivanov, Phys. Rev. Lett. **101**, 103001 (2008)
49. P.H. Bucksbaum, A. Zavriyev, H.G. Muller, D.W. Schumacher, Phys. Rev. Lett. **64**, 1883 (1990)
50. M. Drescher, M. Hentschel, R. Kienberger, G. Tempea, C. Spielmann, G.A. Reider, P.B. Corkum, F. Krausz, Science **291**, 1923 (2001)
51. J. Mauritsson, P. Johnsson, E. Mansten, M. Swoboda, T. Ruchon, A. LHuillier, K.J. Schafer, Phys. Rev. Lett. **100**, 073003 (2008)
52. P. Johnsson, J. Mauritsson, T. Remetter, A. LHuillier, K.J. Schafer, Phys. Rev. Lett. **99**, 233001 (2007)
53. K.P. Singh, F. He, P. Ranitovic, W. Cao, S. De, D. Ray, S. Chen, U. Thumm, A. Becker, M.M. Murnane, H.C. Kapteyn, I.V. Litvinyuk, C.L. Cocke, Phys. Rev. Lett. **104**, 023001 (2010)
54. P. Ranitovic, X.M. Tong, B. Gramkow, S. De, B. DePaola, K.P. Singh, W. Cao, M. Magrakvelidze, D. Ray, I.A. Bocharova, H. Mashiko, A.S. Sandhu, E. Gagnon, M.M. Murnane, H.C. Kapteyn, I.V. Litvinyuk, C.L. Cocke, New J. Phys. **12**, 013008 (2010)
55. P. Rivière, O. Uhden, U. Saalmann, J. Rost, New J. Phys. **11**, 053011 (2009)
56. P. Rivière, C. Ruiz, J. Rost, Phys. Rev. A **77**, 033421 (2008)
57. F. Morales, J.F. Pérez-Torres, J.L. Sanz-Vicario, F. Martín, Chem. Phys. **366**, 58 (2009)
58. U. Fano, Phys. Rev. **124**, 1866 (1961)
59. H. Feshbach, Ann. Phys. **19**, 287 (1962)

60. F. Martín, J. Phys. B Atom. Mol. Opt. Phys. **32**, R197 (1999)
61. K. Ito, R.I. Hall, M. Ukai, J. Chem. Phys. **104**, 8449 (1996)
62. Z.X. He, J.N. Cutler, S.H. Southworth, L.R. Hughey, J.A.R. Samson, J. Chem. Phys. **103**, 3912 (1995)
63. C.J. Latimer, J. Geddes, M.A. McDonald, N. Kouchi, K.F. Dunn, J. Phys. B Atom. Mol. Opt. Phys. **29**, 6113 (1996)
64. G. Sansone, F. Kelkensberg, J.F. Pérez-Torres, F. Morales, M.F. Kling, W. Siu, O. Ghafur, P. Johnsson, M. Swoboda, E. Benedetti, F. Ferrari, F. Lépine, J.L. Sanz-Vicario, S. Zherebtsov, I. Znakovskaya, A. LHuillier, M.Y. Ivanov, M. Nisoli, F. Martín, M.J.J. Vrakking, Nature **465**, 763 (2010)
65. D. Dowek, J.F. Pérez-Torres, Y.J. Picard, P. Billaud, C. Elkharrat, J.C. Houver, J.L. Sanz-Vicario, F. Martín, Phys. Rev. Lett. **104**, 233003 (2010)
66. G. Sansone, F. Kelkensberg, F. Morales, J. Pérez-Torres, F. Martín, M. Vrakking, IEEE J. Sel. Top. Quant. Electron. **18**, 520 (2012)
67. I. Sánchez, F. Martín, Phys. Rev. Lett. **79**, 1654 (1997)
68. I. Sánchez, F. Martín, Phys. Rev. A **57**, 1006 (1998)
69. F. Kelkensberg, C. Lefebvre, W. Siu, O. Ghafur, O. Atabek, A. Keller, V.V. Serov, P. Johnsson, M. Swoboda, T. Remetter, A.L. Huillier, S. Zherebtsov, G. Sansone, E. Benedetti, F. Ferrari, Phys. Rev. Lett. **103**, 123005 (2009)
70. A. Rudenko, T. Ergler, B. Feuerstein, K. Zrost, C.D. Schröter, R. Moshammer, J. Ullrich, Chem. Phys. **329**, 193 (2006)
71. F. Kelkensberg, W. Siu, J.F. Pérez-Torres, F. Morales, G. Gademann, A. Rouzée, P. Johnsson, M. Lucchini, F. Calegari, J. Sanz-Vicario, F. Martín, M. Vrakking, Phys. Rev. Lett. **107**, 043002 (2011)

Chapter 2
Enhanced Ionization of Molecules in Intense Laser Fields

Andre D Bandrauk and François Légaré

Abstract Molecules exposed to intense ultrashort laser pulses undergo rapid ionization at critical large internuclear distances requiring nonperturbative models to explain the highly nonlinear nonperturbative response. Simple quasistatic models allow for the prediction of these critical internuclear distances. It is shown that the mechanism for enhanced ionization is due to overbarrier ionization of the Stark-shifted LUMO (lowest unoccupied molecular orbital) in both symmetric diatomic and triatomic molecules. In nonsymmetric molecules, permanent dipole moments contribute to the strong Stark-shifts of electron orbitals and enhanced ionization occurs through resonances of the HOMO (highest occupied molecular orbital) and LUMO, corresponding to charge transfer. Double ionization is also enhanced at certain critical distances but involves considerable excitations of intermediate electronic states.

2.1 Introduction

Advances in current laser technology are providing new tools for experimentalists and challenges for theorists to explore a new regime of light–matter interaction, the nonlinear nonperturbative interaction of atoms and molecules with intense ($I \geq 10^{+14}\,\mathrm{W/cm^2}$) few-cycle laser pulses. Much of the earlier experimental and theoretical work in this new nonperturbative regime of radiation–atom interaction in atomic systems has been summarized by Brabec and Krausz [1], Corkum and

A.D. Bandrauk (✉)
Laboratoire de chimie théorique, Université de Sherbrooke, Québec, J1K 2R1, Canada
e-mail: andre.bandrauk@usherbrooke.ca

F. Légaré
Institut National de Recherche Scientifique, Centre Énergie Matériaux et Télécommunications, Varennes, Québec, J3X 1S2, Canada
e-mail: legare@emt.inrs.ca

K. Yamanouchi et al. (eds.), *Progress in Ultrafast Intense Laser Science VIII*, Springer Series in Chemical Physics 103, DOI 10.1007/978-3-642-28726-8_2,

© Springer-Verlag Berlin Heidelberg 2012

Krausz [2]. Thus in the atomic case new nonperturbative optical phenomena have been found such as above-threshold ionization (ATI), [1], tunnelling ionization predicted as early as 1980 [3–5] and is the basis of recollision physics [6] which has been confirmed experimentally recently [7]. Another important nonlinear process, high-order harmonic generation (HHG), is the current most convenient source of attosecond ($1\,asec = 10^{-18}$ s) pulses [8]. This rapid development of ultrashort intense laser pulses allows nowadays for shaping and focussing such pulses to higher intensities creating electric fields E greater than the atomic unit, a.u., $E_0 = 5 \times 10^9\,\mathrm{V\,cm^{-1}}$ at the 1s orbit radius $a_0 = 0.0529\,\mathrm{nm}$ of the H atom. This corresponds to an a.u. of intensity $I_0 = cE_0^2/8\pi = 3.54 \times 10^{16} I_0 = cE_0^2/8\pi = 3.54 \times 10^{16}\,\mathrm{W/cm^2}$. Such pulses can be compressed to times of several femtoseconds ($1\,\mathrm{fs} = 10^{-15}$ s), thus creating radiative coherences that are shorter than molecular nonradiative (radiationless) relaxation times [9, 10]. The Schwinger limit, $10^{29}\,\mathrm{W/cm^2}$, is the limit of instability of matter by tunneling from the vacuum itself thus creating electron–positron pairs [11], and electron–positron pairs can now be created from vacuum.at smaller intensities, $I \sim 10^{26}\,\mathrm{W/cm^2}$ [12].

Whereas at the end of the twentieth century the focus was on femtosecond (fs) photochemistry and photophysics culminating with a Nobel prize to A. H. Zewail (Caltech) for "femto chemistry" [13], a major effort is now underway to develop single attosecond (asec) optical pulses based on the nonperturbative physics of HHG in atoms [8] and molecular high order harmonic generation (MHOHG) in molecules [14]. These new asec pulses presage the control of electron motion in molecules, the "Holy Grail" of many atomic and molecular sciences from chemistry to biology to material sciences [15–17].

The intensities discussed in this chapter, $10^{14} \leq I \leq 10^{15}\,\mathrm{W/cm^2}$ correspond to fields approaching the internal Coulomb potentials V_0 or corresponding electric fields E_0 (Table 2.1) in atoms and molecules, thus introducing considerable distortions of intermolecular potentials. Using a dressed photon state representation, such strong radiatively induced distortions create, laser induced molecular potentials (LIMP'S), leading to "bond-softening" via laser-induced avoided crossings of molecular potentials [18, 19]. At these high intensities, one needs to consider further ionization and above-threshold dissociation (ATD), the equivalence of ATI in atoms. Furthermore, the quasistatic picture of atomic tunneling ionization [3,5,6] needs to be modified in view of the multi-centre nature of electron potentials in molecules and the presence of large radiative transition moments, originally described by Mulliken [20] as charge resonance (CR) transitions [18, 19]. One of the fundamental differences between intense field ionization of atoms and molecules is "enhanced ionization" which occurs at critical internuclear distances R_c greater than equilibrium distances R_e for molecules exposed to intense laser pulses [21–24]. Quasistatic models of enhanced ionization have been successful in predicting values of R_c. The main tenet of this model, which differs fundamentally from quasistatic atomic ionization, is that laser-induced charge resonance (CR) effects localize temporarily the electron by charge transfer [22–24] due to the presence of large electronic transition moments in molecules resulting in adiabatic electronic charge transfer across the whole length of a molecule.

Table 2.1 Atomic units ($e = \hbar = m_e = 1$; $c = 137.036$)

Potential energy	$V_0 = e^2/a_0 = 1$ Hartree $= 27.2\,\text{eV}$
Electric field	$E_0 = e/a_0^2 = 5.14 \times 10^9\,\text{V cm}^{-1}$
Intensity	$I_0 = cE_0^2/8\pi = 3.54 \times 10^{16}\,\text{W cm}^{-2}$
Distance	$a_0 = 0.0529\,\text{nm}$
Velocity	$\upsilon_0 = 2.19 \times 10^8\,\text{cm s}^{-1}$
Time	$t_0 = a_0/\upsilon_o = 24.2 \times 10^{-18}\,\text{s} = 24.2$ attoseconds
	$t_c = \hbar/m_e c^2 = 1.29 \times 10^{-21}\,\text{s} = 1.29$ yoctoseconds

2.2 Quasistatic Models: Diatomic One-Electron Systems in Strong Fields

The first simple analytic formulae for atomic multiphoton ionization beyond usual perturbation (Fermi Golden Rule) theory were obtained by Keldysh [3] who showed that a parameter γ, today called the Keldysh parameter [25, 26], allows to separate the perturbative multiphoton regime from the nonperturbative quasistatic tunnelling regime. This parameter is obtained from the low frequency limit of the transition probability from an initial bound state to a Volkov state, the state of a free electron "dressed" by a laser field [25]. This is essentially a time-dependent distorted wave born approximation (DWBA), where the Coulomb potential is retained in the initial bound state and the laser field only in the final continuum state. The Keldysh parameter is then defined by

$$\gamma = (I_p/2U_p)^{1/2}, \tag{2.1}$$

where I_p is the ionization potential and U_p is the ponderomotive energy, $U_p = I_0/4\omega^2$, the average kinetic energy of a free electron in a linearly polarized field $E(t) = E_0\cos(\omega t)$ and peak intensity $I_0 = cE_0^2/8\pi$ at frequency ω. The ponderomotive energy U_p and ponderomotive radius α_p are in fact two additional important parameters in laser-induced electronic processes. Solving the classical equations of motion in a monochromatic time-dependent field $E(t) = E_0\cos(\omega t + \phi)$ gives in a.u. (Table 2.1)

$$\ddot{z} = -E_0\cos(\omega t + \varphi), \dot{z}(t) = -\frac{E_0}{\omega}[\sin(\omega t + \varphi) - \sin\varphi] \tag{2.2}$$

$$z(t) = -\frac{E_0}{\omega^2}[\cos\varphi - \cos(\omega t + \varphi) - \omega t \sin\varphi], \tag{2.3}$$

where ϕ is the initial laser phase called the CEP (carrier envelope phase) at which moment an electron is ionized at time $t = 0$ assuming an initial zero velocity $\dot{z}(0) = 0$ as in the tunnelling model [4], from which one can estimate initial ionization rates using a static E field rate [3, 4, 25]. Tunneling ionization has now been detected in atoms from the modulation of the ionized electron density by the intense laser field [7]. These classical equations define the ponderomotive radius α_p and ponderomotive energy U_p (in a.u.)

$$\alpha_p = E_0/\omega^2,\ U_p = \frac{1}{4}\alpha_p^2\omega^2 = E_0^2/4\omega^2 \qquad (2.4)$$

The Keldysh parameter γ, (2.1) is a ratio of two energies: I_p the minimum energy to ionize the electron and $2U_p$, the maximum kinetic energy, $\frac{1}{2}\dot{z}^2$, acquired by a free electron in the laser field. This simple classical quasistatic model described by (2.2)–(2.4) where one assumes an electron is ionized with initial zero velocity $\dot{z}(t_0) = 0$, the basis of tunnelling ionization models [1, 4], allows us to deduce the laser-induced dynamics of the electron after ionization. Thus for $\lambda = 1{,}064\,\text{nm}$ ($\omega = 0.043$ a.u) and intensity $I = 1 \times 10^{14}\,\text{W/cm}^2 = 3 \times 10^{-3}$a.u., one obtains $\alpha_p = 29\,\text{a.u.} = 1.53\,\text{nm}$. This maximum displacement α_p determines the minimum size of numerical grids [26] and U_p determines the corresponding spatio-temporal steps Δx, Δt (from uncertainty principle relations) in computations at high intensities. Equation (2.2) also allows predicting the maximum average kinetic energy to be $\frac{1}{2}\langle\dot{z}(t)\rangle^2 = 3U_p$ at CEP phase $\varphi = \pi/2$ ($\langle\cos^2\omega t\rangle = 1/2$, $\langle\cos\omega t\rangle = 0$). Such high energies acquired during ionization was observed and explained in microwave ATI using the simple model of (2.2)–(2.3) [27]. In practice, ionization occurs at different phases $\phi = \omega t_0$, i.e., at time t_0, the instant the electron is created in the field. As we show later, t_0, determines the electron trajectories for collision with neighbouring ions and recollision with the parent ion in both linear and circularly polarized laser fields. Initial *zero* velocity, $\dot{z}(t_0) = 0$ will occur at the instant of tunnel ionization [4–6]. The model can also be extended to *nonzero* initial velocity, $\dot{z}(t_0) > 0$ as in preionization with X-ray-VUV pulses [28] and to circularly polarized light [29].

The quasistatic model allows further for establishing the critical or minimum field E_m where above-barrier ionization occurs instead of under-barrier tunnelling ionization [21–24]. This is illustrated in Fig. 2.1a for a one-electron atom and Fig. 2.1b for a single valence electron in a diatomic molecule such as $H_2{}^+$, in the presence of a static electric field E. In the atomic case, where a single "active" electron is bound by an effective nuclear charge $q+$, the total potential along the field polarization, i.e., z-axis, can be written as $V(z) = -q/|z| - Ez$. As seen from Fig. 2.1a, the electric field distorts the atomic Coulomb potential thus producing a barrier with a maximum at $z_m = (q/E_m)^{1/2}$ obtained from $\partial V/\partial z = 0$. Setting $V(z_m) = -I_p$, where I_p is the ionization potential, one obtains the minimum field $E_m = I_p^2/4q$ for above-barrier ionization. For hydrogenic levels, $I_p = -\varepsilon_n = q^2/2n^2$, for a level of principal quantum number n. For such a level, the minimum electric field E_m for atomic above-barrier ionization is $E_m = q^3/(2n)^4$ a.u. or intensity $I_m = cE_m^2/8\pi = cq^6/(8\pi(2n)^8)$ a.u. Thus for the $n = 1$ (1s) level of H, $I_p = 1/2$ a.u., $I_m = 1.4 \times 10^{14}\,\text{W/cm}^2$. For the Th^{+89} ion in its $n = 2$ level, $I_m = 90^6/2^{16}$ a.u. $= 2.84 \times 10^{23}\,\text{W/cm}^2$. Such superintense fields are being currently considered in the European ELI (extreme light infrastructure) project [30]. The theoretical description of super intense field electron ionization

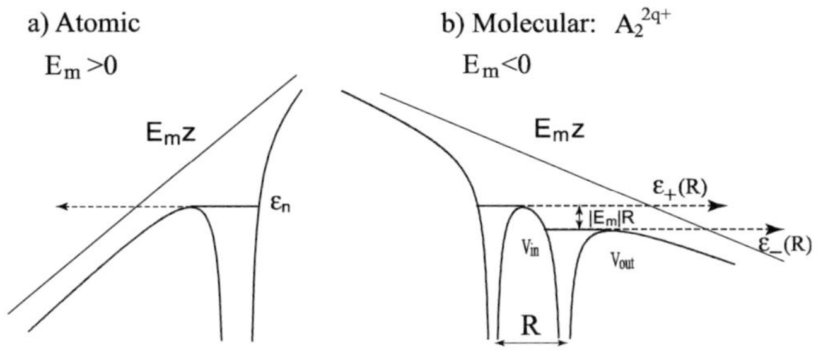

Fig. 2.1 Above barrier ionization in (**a**) atoms and (**b**) molecules via Stark-shifted LUMO, $\varepsilon_+(R)$ and HOMO, $\varepsilon_-(R)$

requires applying the relativistic Dirac equation to such processes and is becoming a new active field of research [31–33].

Molecular above-barrier ionization leads to a new concept, CREI (charge resonance enhanced ionization), for which recent experiments are summarized in [21]. As illustrated in Fig. 2.1b, a molecular electronic potential is a multicenter potential. Atomic energy levels are transformed into delocalized molecular orbitals, MO's, which at large internuclear distance R become linear combinations of atomic orbitals. In the case of H_2^+, the two most important MO's in low frequency multiphoton processes are the HOMO, highest occupied MO and LUMO, lowest unoccupied MO, as these couple radiatively through an electronic transition moment $<\text{HOMO}|z|\text{LUMO}>= R/2$ [19, 20]. In the presence of a static field, the LUMO is Stark-shifted in energy by $+ER/2$ whereas the HOMO's energy is lowered by $-ER/2$. Thus, at some critical internuclear distance R_c and critical field E_m, the LUMO, which can be populated nonadiabatically by the field [34, 35], ionizes directly over the barrier between the two nuclei. This simple quasistatic model was first proposed by Codling et al. [36–38] neglecting the Stark energies of both HOMO and LUMO, but rather emphasizing the localization (suppression of the tunnelling) between the field free wells, (Fig. 2.1b). Inclusion of the Stark energy shift of the LUMO allows for analytical estimates of R_c and E_m for above-barrier ionization in symmetric diatomic molecules [23, 39]. Inclusion of static dipole moments such as in HeH^{2+} further generalizes the concept of CREI in nonsymmetric molecules [40].

We follow the original 1-D model of one-electron molecular ionization in ${A_2}^{2q+}$ systems, (Fig. 2.2b), where $q+$ is the charge on each nucleus [23, 39]. Exact Born–Oppenheimer (static nuclei) time-dependent Schrödinger, TDSE, simulations of highly charged diatomic molecular ions exhibit ionization maxima for internuclear distances $6 \leq R \leq 10$ a.u. at "critical" distances R_c for laser polarization parallel to the internuclear axis. Ionization from MO's with densities perpendicular to the intermolecular axis also induces abrupt increase of ionization into atomic plateaus at $R_c \sim 8$ a.u. [40, 41]. The necessary conditions for CREI to occur at R_c is that the upper Stark shifted electronic eigenstate energy $\varepsilon_+(R)$ exceeds the two potential barriers in Fig. 2.1b, i.e.

$$V_{\mathrm{out}}(R_{\mathrm{c}}) \leq V_{\mathrm{in}}(R_{\mathrm{c}}) \leq \varepsilon_{+}(R_{\mathrm{c}}), \tag{2.5}$$

where V_{in} and V_{out} are the inner and outer barriers of field modified the two-center Coulomb potential at the critical field strength E_m,

$$V(z, R) = \frac{-q}{\left|\frac{R}{2} + z\right|} - \frac{q}{\left|\frac{R}{2} - z\right|} + E_m z, \tag{2.6}$$

and

$$\varepsilon_{+}(R) = -qI_p - q/R + E_m R/2. \tag{2.7}$$

$E_m R/2$ is the energy Stark shift of the LUMO (Fig. 2.1b), I_p is the ionization potential of the neutral atom and $\frac{-q}{R}$ is the electron Coulomb attraction with a neighbouring ion. Ionization potentials of electrons from the same shell are assumed to scale as $I(\mathrm{ion}) = qI_p$ for an ion of charge q–1. The potential (2.6) has maxima at z_{in} and z_{out} for $E_m < 0$ as in Fig. 2.1b which are given by [23],

$$z_{\mathrm{out}} = -\left(R/2 + E_q^{-1/2}\right),\; z_{\mathrm{in}} = E_q R^3/32 \tag{2.8}$$

where $E_q = |E_m|/q$.

Solution of the CREI conditions (2.5) by setting $\varepsilon_{+}(R_{\mathrm{c}}) = V_{\mathrm{out}}(R_{\mathrm{c}}) = V_{\mathrm{in}}(R_{\mathrm{c}})$ gives [23, 39],

$$R_{\mathrm{c}} = 4.07/I_p,\; |E_m| = 0.139 I_p^2 \tag{2.9}$$

Neglecting the Stark-shift of the LUMO gives the value of $R_{\mathrm{c}} = 3/I_p$ and defines the onset of electron localization due to negligible electron tunnelling between the potential double wells in the unperturbed field-free molecule [23,24]. Equation (2.8) shows the importance of the static field E_m in Stark-shifting the LUMO energy to $\varepsilon_{+}(R)$ above the field modified barriers illustrated in (Fig. 2.1b). Nevertheless, the end result is the surprising independence of R_{c} from field strengths E_m and nuclear charge $q+$, thus establishing CREI as a *universal* intense field nonperturbative phenomenon [21,34–60]. A simple physical explanation of CREI also emerges from the quasistatic image, Fig. 2.1b. The ionizing electron in the Stark-shifted LUMO of energy $\varepsilon_{+}(R)$ is initially accelerated by the neighboring ion whereas the electron in the HOMO, $\varepsilon_{-}(R)$, is initially decelerated by its parent ion as it leaves it, thus making ionization from the LUMO the dominant mechanism for CREI.

Nonsymmetric molecules such as the one-electron HeH^{2+} [47, 48] present a different nonlinear nonperturbative response due to the presence of a permanent dipole moment. This is illustrated in Fig. 2.2a for $E_m > 0$ and Fig. 2.2b for $E_m < 0$. Thus, the HOMO is displaced upward (↑) to $\varepsilon_{+}(R)$ by $|E_m|R/2$ for $E_m < 0$ (Fig. 2.2b) and becomes resonant (degenerate) with the LUMO, $\varepsilon_{-}(R)$ which is shifted downward (↓). Thus, enhanced ionization, EI, can occur by resonance between these two levels. In the case of the opposite positive field, $E_m > 0$ (Fig. 2.2a), both HOMO and LUMO can never cross, thus suppressing the resonance and enhanced ionization. This qualitative picture of CREI in symmetric molecules

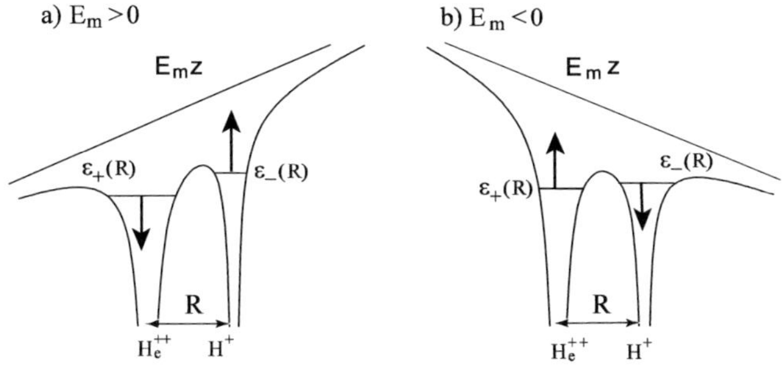

Fig. 2.2 Enhanced ionization in HeH^{2+} by Stark-shifted ε_+ (LUMO), ε_- (HOMO)

illustrated in Fig. 2.2 emphasizes the importance of resonances in these molecules with nonvanishing dipole moment. In fact, an avoided crossing between the HOMO and LUMO occurs for $E_m > 0$ (see Fig. 10 in [47]), inducing a new harmonic generation spectrum at high intensities [48]. Using the similar static Stark-displaced potential model as in the symmetric diatomic case above, one readily obtains the critical distance R_c for enhanced ionization in nonsymmetric systems [47] to occur for $E_m < 0$ at

$$R_c = \frac{(I_1 - I_2) + [(I_1 - I_2) - 4|E_m|(Z_1 - Z_2)]^{1/2}}{2|E_m|}. \tag{2.10}$$

where I_1 and I_2 are the ionization potentials of the different heteroatoms with $I_1 > I_2$ and effective nuclear charges Z_1, Z_2. Numerical TDSE simulations for HeH^{2+} where $Z_1 = 2$ and $Z_1 = 1$ show for this nonsymmetric system dependence of R_c on the CEP phase φ of the field strength E_m and charge separation Z_1–Z_2 as predicted by (2.10). Clearly in this nonsymmetric case, the critical distance R_c for enhanced ionization depends on the CEP, ϕ, the field amplitude $|E_m|$, and the I_p difference between the heteroatoms, thus offering the possibility of controlling this highly nonlinear nonperturbative electron response in nonsymmetric molecules. The simple one-barrier models of enhanced ionization illustrated in Figs. 2.1–2.2 corresponding to quasistatic (electrostatic) models are confirmed by exact 3-D calculation of ionization rates using time-dependent Schrödinger equations, TDSE's for one-electron molecules $H_2{}^+$ [41–43] and HeH^{2+} [47]. In Fig. 2.3 we illustrate the ionization rate for $H_2{}^+$ as a function of internuclear distance R (a.u.) at intensity $I = 2 \times 10^{14}\,\mathrm{W/cm^2}$ for a linearly polarized pulse, wavelength $\lambda = 800$ um ($\omega_L = 0.0567$ a.u. $\tau_L = 2.67$ fs) pulse duration five cycles. The circularly polarized pulse is in the (x, y) plane of the molecule perpendicular to the internuclear (z) R axis. The linearly polarized pulse has its electric field parallel or perpendicular to the (z) R axis. The parallel linearly polarized pulse results (+) show two maxima at $\lambda = 800\,\mu$m with ionization rates $\sim 1 \times 10^{13}\,\mathrm{s}^{-1}$. Since I_p (H(1s)) $= 0.5$ a.u., (2.9) predicts $R_c \sim 8$ a.u. The first peak at $R \simeq 6$ a.u. in Fig. 2.3 a.u. corresponds to a photon resonance at energy $\sim$ 10 eV for the ground σ_g (1s) to excited σ_u (1s)

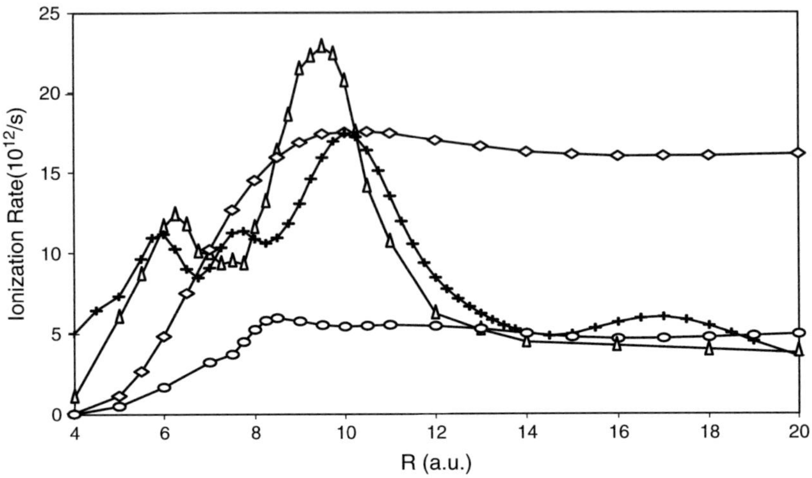

Fig. 2.3 Ionization rates for H_2^+ vs R for (i) $I = 10^{14}\,\mathrm{W/cm^2}$, linear polarization parallel to R (z-axis): (**a**) (*plus*): $\lambda = 800\,\mathrm{nm}$; (**b**) (*open triangle*): $\lambda = 1,064\,\mathrm{nm}$, and (**c**) (*open circle*): $\lambda = 800\,\mathrm{nm}$, perpendicular to R and (ii) $I = 2 \times 10^{14}\,\mathrm{W/cm^2}$, circular polarization in x–y plane, $\lambda = 800\,\mathrm{nm}$: (open diamond)

CR transition in H_2^+ [39–42]. Clearly, the region $R \geq 8\,\mathrm{a.u.}$ is dominated by atomic transitions for perpendicular linear and circularly (◇) polarized pulses which are constant for large R. For parallel linearly polarized pulses the double CREI maximum at $8 \leq R \leq 10\,\mathrm{a.u.}$ at $\lambda = 800\,\mathrm{nm}$ (+) becomes a single maximum at longer wavelength $\lambda = 1,064\,\mathrm{nm}$ (Δ) thus eliminating multiphoton resonances and rendering the static model of (2.8)–(2.9) more rigorous. Thus in Fig. 2.3 clear maxima at $R = 6\,\mathrm{a.u.}$ with ionization rate $\sim 10^{13}\,\mathrm{s^{-1}}$ and $R = 9 - 10\,\mathrm{a.u.}$ with rate $2 \times 10^{13}\,\mathrm{s^{-1}}$ correspond to lifetimes of $\leq 100\,\mathrm{fs}$ with $.I'_p.\mathrm{s} = 0.5 + \frac{1}{R} \sim 0.7 - 0.6\,\mathrm{a.u.}$ ($\sim$ 20–16 eV's)

The nonsymmetric HeH^{2+} one-electron molecular system contrary to the symmetric H_2^+ system is influenced strongly by a permanent dipole moment in the ground state configuration He^+ (1s) H^+ which changes polarity in the first excited state He^{++} H (1s) following charge transfer [47, 48]. The resulting ionization probabilities are illustrated in Fig. 2.4 at intensities $I = 10^{15}$, 3×10^{15}, and $5 \times 10^{15}\,\mathrm{W/cm^2}$ for CEP's $\phi = 0$ and π, i.e., for $E_m > 0$ and < 0, respectively. As predicted by (2.10) and illustrated in Fig. 2.2, only for negative field $E_m < 0$, $\phi = \pi$, the HOMO (He^+(1s)) and LUMO (H(1s)) cross, resulting in enhanced ionization at $4 \leq R_c \leq 6\,\mathrm{a.u.}$ for intensities $5 \times 10^{15} \geq \mathrm{I} \geq 10^{15}\,\mathrm{W/cm^2}$. Such field-amplitude-dependent ionization rates demonstrate the importance of laser control of charge transfer, CR, at high intensities in nonsymmetric molecules. Enhanced ionization predicted theoretically [22–24] has been confirmed experimentally [21], in diatomics and complex molecules [56]. Furthermore, such a quasistatic model of Coulomb explosion of clusters [57] suggests that enhanced ionization may be

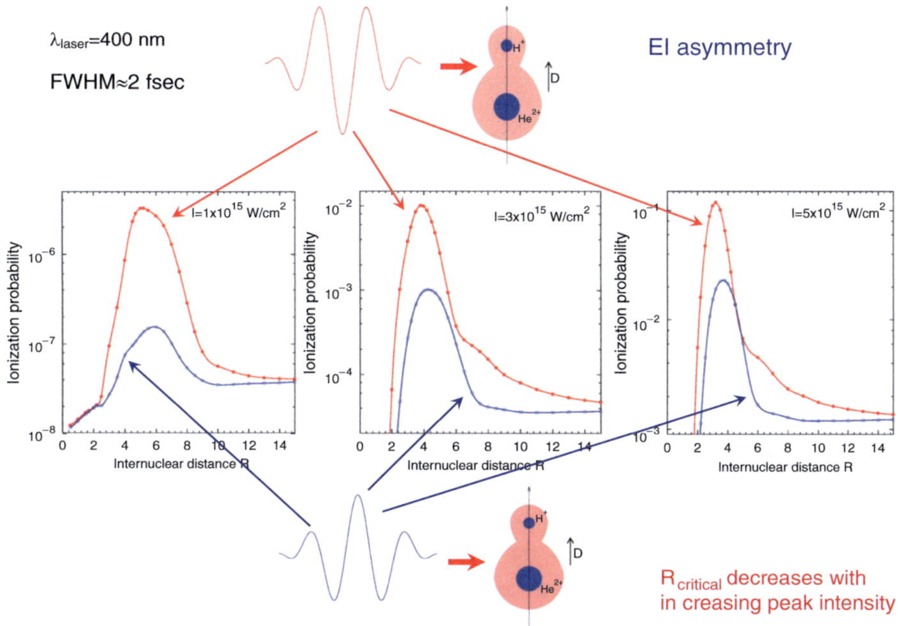

Fig. 2.4 HeH^{2+} ionization probabilities for: fields $E(\tau, \phi)$ and CEP $\phi = 0$ (*blue*), π (*red*) at peak intensities $I = 10^{15}$, 3×10^{15}, and 5×10^{15} W/cm^2, as a function of internuclear distance R with maxima at R_c (R_{critical})

a mechanism for laser fusion [58, 59]. In nonsymmetric molecules, the strong orientation and CEP dependence of enhanced ionization as shown in simulations of one-electron nonsymmetric molecular systems suggest applications in the control of ionization of complex molecules [60], in the laser-assisted population inversion in heavy ion collisions for X-ray laser production [61] and finally in the imprinting of nanostructures in heteroatomic lattices with ultrashort intense few-cycle pulses [62]. With the extension of the CREI model to one-electron nonlinear molecules such as $H_3{}^{2+}$ [63], multi-center systems [64] have predicted this nonlinear nonperturbative phenomenon to be unique to molecules, i.e., over-barrier ionization results in complete ionization at critical distances R_c. Nuclear motion of $H_2{}^+$ for which the vibrational period $\tau_\nu \simeq 10$ fs has the tendency to "wash out" the double peak structure of the ionization shown in Fig. 2.3 [65, 66]. Exact non-Born-Oppenheimer simulations of one-dimensional one-electron model systems $H_2{}^+$ and $H_3{}^{2+}$ with quantum moving nuclei show that coherent excitations of the HOMO and LUMO occur during the nuclear motion at the critical distances R_c [67]. Such coherent efforts during CREI resulting in two-surface electron population dynamics have been observed recently by *asec* strobing [68] and modeled successfully with non-Born-Oppenheimer electron—nuclear TDSE's [69]. Such non-adiabatic simulations have shown highly correlated electron-nuclear motion in dissociating $H_2{}^+$ [70] resulting in multiple ionization bursts on *asec* time scale due to strong field-modulated diffraction effects [71] due to laser-induced electron diffraction, LIED [72].

2.3 Double Ionization in Strong Fields

In Sect. 2.2, we have described quasistatic above-barrier simple electron ionization in one-electron molecules for which analytic models were obtained to derive expressions for the critical distance R_c for CREI in symmetric molecules and enhanced ionization in nonsymmetric molecules. Double ionization introduces a new phenomenon—nonsequential double ionization, NSDI, observed as early as 1992 with optical tunnelling of He [73] and explained based on the 3-step model of tunneling ionization, where electron propagation in an intense laser pulse is followed by recollision with maximum energy $3.17\,U_p$ [4]. Tunneling ionization imposes the initial velocity condition $\dot{z}(t_0) = 0$ at time t_0. This model has been generalized to nonzero initial velocity $\dot{z}(t_0) \neq 0$ where it was shown that the return collision energy E_c never exceeds 3.17 U_p i.e., $E_c \leq 3.17\ U_p$ [28]. Collision with neighboring ions can, however, lead to energies exceeding $E_c > 3.17\ U_p$ and is R dependent [57, 74–76]. In such processes, charge transfer or ionic states, such as $H^+\ H^-$ in H_2, $O^+\ O^-$ in O_2, produce very large transition moments emphasized early by Mulliken [20]. Ab initio calculations of H_2 in strong fields have proven the existence of $H^+\ H^-$ as a precursor to enhanced ionization [77–79] as well as in the triatomic $H_3{}^+$ in its linear [80] and nonlinear [81] geometry. 1-D non-Born-Oppenheimer simulations of H_2 have also identified the CR state H^+H^- as an important doorway state in the Coulomb explosion of this system [82]. The presence of a pair of electrons implies considerable electron correlation in the nonlinear nonperturbative response of molecules to intense ultrashort pulses [83–85]. However in molecules, charge transfer in ionic states such as $H^+\ H^-$ and $O^+\ O^-$ as recently suggested in O_2 time-resolved dissociative ionization [86] implies dipole moments which will introduce large Stark-shifts in the molecular potentials at high laser intensities. It was found earlier that the dipole moment of asymmetric charge distributions interacts with intense external fields leading to charge asymmetric dissociation of highly charged ions [87]. Ionization from these asymmetric charge distributions can lead to shake up excitation of the remaining bound electrons [88] by interaction with the ionized electron. CEP sensitivity of these charge transfer, CR, processes has been now confirmed in dissociative ionization of CO [89], and correlated two-electron ab initio calculations in H_2 show evidence of enhanced excitation in single and double ionization at the CREI critical distance R_c [84]. Ionic states act therefore as doorway states for intense laser-field-enhanced ionization of two-electron molecules. For H_2, one must consider at least the two electronic states, $1\sigma_g{}^2(\mathrm{X}^1\Sigma_g{}^+)$, $1\sigma_g 1\sigma_u(\mathrm{B}^1\Sigma_u{}^+)$ and $1\sigma_u{}^2(\mathrm{E,F}^1\Sigma_q{}^+)$ electronic configurations [77–79, 84]. The asymptotic dissociation products ϕ_1, ϕ_2, and ϕ_3 for these states are expressed in terms of 1s atomic orbitals a and b on each H atom [90],

$$\phi_1(1,2) = [a(1)\,a(2) + b(1)\,b(2) + a(1)\,b(2) + b(1)\,a(2)]/2, \tag{2.11}$$

$$\phi_2(1,2) = [a(1)\,a(2) - b(1)\,b(2)]/\sqrt{2} \tag{2.12}$$

$$\phi_3(1,2) = [a(1)\,a(2) + b(1)\,b(2) - a(1)\,b(2) - b(1)\,a(2)]/2 \tag{2.13}$$

Both ground ϕ_1 and excited ϕ_3 states are linear combinations of *ionic* (*aa*, *bb*) and *covalent* (*ab*) configurations whereas ϕ_2 is purely ionic. The electronic transition moments between these states are thus coupled by the same radiative matrix element ER/$\sqrt{2}$ for a field strength E (intensity $I = cE^2/8\pi$). Diagonalization of the 3×3 energy matrix yields readily the *adiabatic* energies and states [78]

$$\varepsilon_0 = 0,\ \phi_0 = [a(1)\,b(2) + b(1)\,a(2)]/\sqrt{2} \tag{2.14}$$

$$\varepsilon_+ = ER,\ \phi_+ = b(1)\,b(2), \tag{2.15}$$

$$\varepsilon_- = -ER\,\phi_- = a(1)\,a(2), \tag{2.16}$$

The ground state is ϕ_- corresponding to the ionic state $H_a^-\ H_b^+$ whose energy agrees with the electrostatic energy of a charge displaced (transferred) through the distance R by a field of amplitude E. The ionic states ϕ_-, ϕ_+ acquire additional Coulomb energy. Thus, if we consider the general charge transfer $A^{q+}\ A^{q+}\ A^{(q-1)+}\ A^{(q+1)+}$, the initial total energy is

$$\varepsilon_{q,q}(R) = -2I_p(q) + q^2/R, \tag{2.17}$$

where $I_p(q)$ is the ionization potential of the ion A^{q+} and q^2/R is the ion–ion repulsion. The final energy of the charge transfer state in the field E is

$$\varepsilon_{q-1,q+1}(R) = -I_P(q-1) - I_P(q) + (q+1)(q-1)/R - ER, \tag{2.18}$$

and the energy difference is

$$\Delta\varepsilon(R) = \Delta I_p - 1/R - ER, \tag{2.19}$$

where $\Delta I_p = I_p(q) - I_p(q{-}1)$. The critical field strength E_c required for the crossing of the covalent $A^{q+}\ A^{q+}$ and ionic $A^{(q-1)+}\ A^{(q+1)+}$ state at critical distance R_c is determined from $\Delta\varepsilon(R_c) = 0$

$$E_c = (\Delta I_p - 1/R_c)/R_c. \tag{2.20}$$

Solutions of the TDSE for H_2 in 1-D [78], and 3-D [84] show in fact that the ionic state $H_a^-\ H_b^+$ due to its large dipole moment indeed becomes the ground state for fields $E > E_c$ and that the ionic state created determines the ionization process. In H_2 since $\Delta I_p = I_p(H) - I_p(H^-) = 0.47$ a.u., one obtains $R_c = 5$ a.u. at intensity $I = 10^{14}\,W/cm^2$ ($E = 0.053$ a.u.) in agreement with 3-D TDSE's [84]. We note that a similar model of charge transfer at critical distance R_c for critical fields E_m

applies to nonsymmetric molecules and is derived in [46] giving (2.10). The charge transfer model for H_2, (2.11)–(2.20), has been applied to linear ${H_3}^{2+}$, ${H_4}^{3+}$, ${H_4}^{2+}$ [61], ${H_3}^{+}$ with static [80] and ${H_3}^{2+}$ [67] and H_2 [91] with moving (non-Born-Oppenheimer) nuclei. In the latter non-Born-Oppenheimer simulations, coherent excitations of the HOMO and LUMO in the CREI region occur and influence both the ATI spectra as well as the Coulomb explosion proton kinetic energies. The presence of such charge asymmetric states has been shown to occur especially for highly charged states at high intensities, $I \geq 5 \times 10^{14}\,\mathrm{W/cm^2}$ in I_2 leading to both charge symmetric and asymmetric dissociation products [87].

Non-symmetric molecules as discussed for the one-electron HeH^{2+} have permanent dipole moments [47, 48] with the effect that enhanced ionization is CEP sensitive as illustrated in Figs. 2.2 and 2.4. This effect persists in the two-electron case [92], HeH^+ with the two configurations He–H^+ and He^+–H with dipole moments $4R/5$ and $-R/5$, respectively. In Fig. 2.5a we report the total single, P_{SI} and double, P_{DI}, ionization probabilities at intensity $I = 10^{15}\,\mathrm{W/cm^2}$ and wavelength $\lambda = 800\,\mathrm{nm}$ ($\omega = 0.057\,\mathrm{a.u.}$), obtained by projection on all bound electronic states. Figure 2.5b reports the accompanying single, P_{SE} and double P_{DE}, electron excitations obtained by projecting the exact 6-D two-electron wavefunction on appropriate field-free electronic states of the HeH^+ molecule [92]. Figure 2.5 shows clearly enhancement of one-and two-electron ionizations and excitations at a critical distance $3 \leq R_c \leq 4\,\mathrm{a.u.}$ We now use as in previous sections a simple quasistatic one-electron approximation model to derive an analytical expression for R_c of HeH^+ ionization in intense fields. At large R, the field-free one-electron ground state energy level of HeH^+ becomes $\varepsilon_1 = -I_p(\mathrm{H})–Z_1/R$ corresponding to He–H^+ configurations [93]. The first excited state He^+–H has energy $\varepsilon_2 = -I_p(\mathrm{H})–Z_1/R$. In the presence of a negative electric field, $|E|$, ε_1 and ε_2 are Stark-shifted via the dipole moments of HeH^+ and He^+H respectively:

$$\varepsilon_1 = -I_p(\mathrm{He}) - Z_2/R + (4R/5)|E_m| \tag{2.21}$$

$$\varepsilon_2 = -I_p(\mathrm{H}) - Z_1/R - (R/5)|E_m|. \tag{2.22}$$

Setting $Z_1 = Z_2 = 1$ corresponds to the nuclear charges of He^+ and H^+. Crossing of the two electronic states with energies ε_1 and ε_2 occurs for negative E_m since I_p (He) $>$ I_p (H), resulting in efficient over-barrier ionization as in HeH^{2+} [47]. This crossing defines the critical distance for enhanced ionization

$$R_c = (I_p(\mathrm{He}) - I_p(\mathrm{H}))/\,|E_m|\,, \tag{2.23}$$

as a function of the field E_m. At intensity $I = 10^{15}\,\mathrm{W/cm^2}$, $E_m = 0.17\,\mathrm{a.u.}$ and $R_c = 3\,\mathrm{a.u.}$ in good agreement with Fig 2.5a [92]. Adding polarizabilities of He and H as suggested recently [94] can be shown to have negligible effect. Equations (2.21)–(2.22) show that it is the permanent dipole moment and the electric field which dominate the nonlinear nonperturbative ionization of asymmetric molecules.

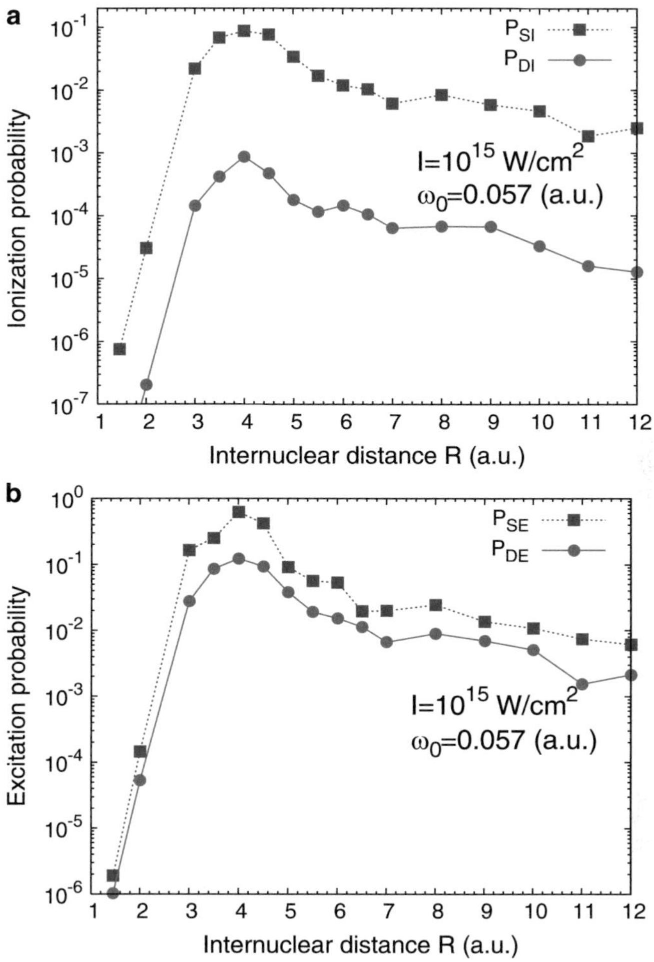

Fig. 2.5 Ionization and excitation probabilities for HeH^+ as a function of internuclear distance R at intensity $I = 10^{15}\,\mathrm{W/cm^2}$, $\omega_0 = 0.057$ a.u. ($\lambda = 800$ nm). (**a**) total single P_{SI} and double P_{DI} ionization; (**b**) total single P_{SE} and double P_{DE} excitation

2.4 Coulomb Explosion Imaging of CREI in Triatomics

Time-resolved Coulomb explosion imaging (CEI) is now an established technique to make molecular movies with sub-Angstrom spatial resolution and sub-5fs temporal resolution [95,96]. CEI of diatomics has shown that molecular bonds stretch towards a critical internuclear distance R_c discussed in the previous sections, where ionization rates are greatly enhanced [21]. This also explains the explosion of large clusters creating highly charged states [57] which can catalyze nuclear fusion [58, 59]. In symmetric diatomics, one-electron systems such as H_2^+/D_2^+, the enhancement of ionization is explained in Sect. 2.2 as CREI in terms of charge resonance, CR, states which are strongly coupled by the laser field at large internuclear distances called critical distances R_c [21–24]. At such internuclear distances, the energy difference between two CR states approaches the incident high intensity photon energy $h\nu$ giving rise to electron *localization* by creating a coherent superposition of opposite parity states [67–71]. For triatomic molecules such as CO_2, the experimental confirmation of a critical geometry for CREI has been recently reported by CEI [97]. Previous experimental results for CO_2 [98, 99] have been compared to over-barrier ionization models discussed for diatomics in previous sections. Numerical simulations of nuclear motion on field dressed molecular potentials such as ${CO_2}^{2+}$ but without ionization have revealed bending and symmetric stretching on a time scale of $\sim$ 100 fs [60]. Using 3-D ion momentum coincidence measurements of the three-body dissociative ionization of CO_2 to ${CO_2}^{n+}$ ($n = 3 - 6$) with few-cycle linearly polarized 800-nm laser pulses, a systematic study of dissociative multiple ionization dynamics has been achieved as a function of pulse duration [97]. This has been supplemented by nonlinear nonperturbative time-dependent density functional theory (TDDFT) calculations [100] allowing for the identification of the relevant CR states and thus the underlying mechanism for CREI in CO_2. It has been previously demonstrated for diatomics that by using few-cycle pulses, CREI can be suppressed since the molecular ion does not have time to reach R_c within the duration of the pulse [101].

Figure 2.6 illustrates the kinetic energy release, KER of the charge state of CO_2 as a function of pulse duration for the $(1,\ 1,\ 1\ \div O^+ + C^+ + O^+)$, $(1,\ 1,\ 2\ \div O^+–C^+–O^{2+})$, $(1, 2,\ 2\ \div O^+–C^{2+}–O^{2+})$, and finally, $(2,\ 2,\ 2\ \div O^{2+}–C^{2+}–O^{2+})$ channels. It is observed that the KER becomes significantly dependent on the pulse duration only for the 4+ and higher charge states. With 7 fs duration and for final charge states higher than 3+, the observed KER is close to the value expected for Coulomb explosion from the CO_2 equilibrium geometry. For longer pulse duration, the total KER impact decreases and becomes almost independent for pulse duration $\tau > 100$ fs [97]. Similar results occur for the KER pulse duration of D_2 with pulse duration > 40 fs [101]. In the D_2 case, the KER from Coulomb explosion becomes independent of pulse duration once the molecular ion stretches to R_c where CREI is known to be operative. Simulation of the Coulomb explosion of the (2, 2, 2) channel using the measured moments and a numerical data inversion algorithm [102] allows to retrieve the molecular structure at 7 fs, ($<R_{CO}> = 1.3$ Å and $<\theta_{OCO}> = 168°$) very close to the equilibrium geometry ($<R_{CO}> = 1.16$ Å and $<\theta_{OCO}>) = 172°$ [102]. For a pulse duration longer than 100 fs, the experimental observations lead to a molecular structure of ($<R_{CO}> = 2.1$ Å and $<\theta_{OCO}> = 163°$) at which CREI occurs.

Since the mechanism for CREI relies on strong radiative interactions between "frontier" orbitals (Sects. 2.2–2.3), HOMO's and LUMO's, nonlinear nonperturbative TDDFT calculations for CO_2 [100] have been used to calculate the ionization rates and relevant electronic transition moments as a function of R_{CO} for intensities $I = 8 \times 10^{14}\,\mathrm{W/cm^2}$, $\lambda = 800$ nm and $\tau = 6$ optical cycles. The numerical results are summarized in Fig. 2.7. In Fig. 2.7a, perpendicular and parallel to

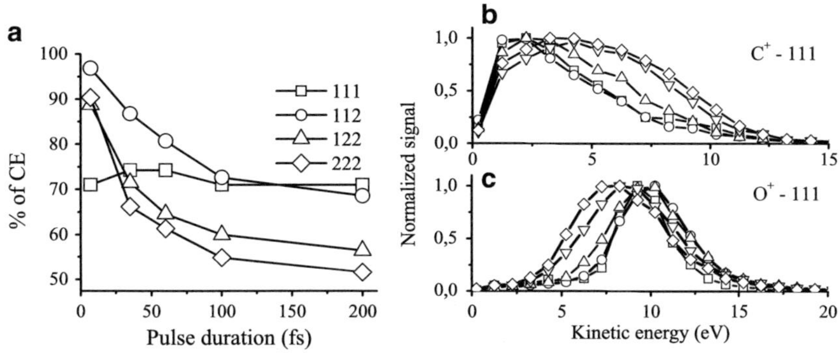

Fig. 2.6 Experimental kinetic energy release, KER, spectra as a function of laser pulse duration: (**a**) ratio of KER vs energy at equilibrium ($R_{CO} = 1.16\,\text{Å}$), (**b**) C^+, and (**c**) O^+: KER for (1,1,1 charge state). Pulse duration and intensity: *open square*—7 fs, 5×10^{14}; *open circle*—35 fs, 5×10^{14}; *open triangle*—60 fs, 3×10^{14}; *inverted triangle*—100 fs, 3×10^{14}; *open diamond*—200 fs, $3 \times 10^{14}\,\text{W/cm}^2$

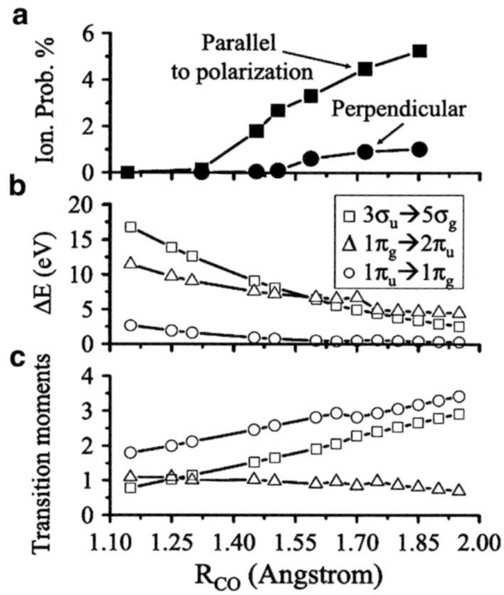

Fig. 2.7 TDDFT simulations (**a**) total ionisation probability for parallel and perpendicular orientations, (**b**) Energy gaps between KS orbitals, (**c**) transition moments (a.u.) for parallel transitions as a function of R_{CO}

laser polarization ionization rates confirm the dominance of parallel ionization. Figure 2.7b shows the energy separation of relevant orbitals as a function of R_{CO}. and Fig. 2.7c reports the corresponding transition moments. Of note is that the $1\pi_u$ (HOMO-2) $\rightarrow$ $1\pi_g$ (HOMO) transition behaves as a CR transition with a transition moment which scales as $R_{CO}/2$ between the antibonding $1\pi_g$ and the inner shell bonding $1\pi_u$. These two orbitals become nearly degenerate at $R_{CO} > 2\,\text{Å}$

with an $R/2$ transition moment, a signature of CREI. Another similar dominant CR transition is the $3\sigma_u$ (HOMO-1) to $5\sigma_g$ (LUMO). The numerical simulations, limited to $R_{CO} \leq 2$ a.u. because of energy degeneracy of orbitals for large $R_{CO.}$, nevertheless show that it is the $3\sigma_u$ orbital which contributes mostly to the ionization as its density is mainly localized along the internuclear axis. Even though the ionization potential I_p of the $3\sigma_u$ orbital is higher than that of the $1\pi_g$, its ionization rate is larger due to its geometry, a general phenomenon at high intensities [100].

To summarize, as the CO_2^{3+} breaks apart through dissociative ionization, the numerical simulations confirm that the ionization rate increases rapidly when the molecule is aligned along the laser polarization axis because of strong, nonperturbative radiative coupling of the main CR states, $3\sigma_u \rightarrow 5\sigma_g$. The increasing ionization rate, linearly varying transition moments and the increasing degeneracy of orbitals as a function of R_{CO} in Fig. 2.7 allow us to characterize the dynamics of the dissociative ionization as due to CREI. This requires CO_2^{3+} to stretch with a time equal to or longer than 100 fs to the critical distance $R_{CO} > 2$ Å. For shorter pulses, CREI is suppressed due to less ionization at shorter distances.

In conclusion, in spite of highly nonlinear, nonperturbative response of molecules to extreme intense electromagnetic pulses, simple classical models have been the guiding models for understanding laser induced electron recollision, LIERC [4, 28, 103] laser imprinting of nanostructures [62], molecular ATI [67], laser induced electron diffraction LIED [72, 104], and charge-resonance-enhanced ionization CREI, reviewed in this chapter. Other examples are HHG [76, 105], time-resolved photoelectron holography [106] and in many cases, controlling the physical processes that occur in this new nonperturbative highly nonlinear regime of laser molecule interaction [107].

Acknowledgements We thank the following colleagues for "illuminating" discussions on strong field molecular physics: T. Brabec, P.B. Corkum, C. Cornaggia, H. Kono, K. Yamanouchi, J. Sanderson, and I. Litvinyuk.

References

1. T. Brabec, F. Krausz, Rev. Mod. Phys. **72**, 545 (2000)
2. P. Corkum, F. Krausz, Nature Phys. **3**, 381 (2007)
3. L. Keldysh, Sov. Phys. JETP **20**, 1307 (1965)
4. P.B. Corkum, Phys. Rev. Lett. **71**, 1994 (1993)
5. P.B. Corkum, N.H. Burnett, F. Brunel, Phys. Rev. Lett. **62**, 1259 (1989)
6. P.B. Corkum, Phys. Today March 36 (2011)
7. A.J. Verhaef et al., Phys. Rev. Lett. **104**, 163904 (2010)
8. F. Krausz, M.Y. Ivanov, Rev. Mod. Phys. **81**, 163 (2009)
9. J.I. Steinfield, *Molecules and Radiation* (MIT Press, Cambridge, USA, 1985)
10. H. Lefebvre-Brion, R.W. Field, *Spectra and Dynamics of Diatomic Molecules* (Elsevier, Holland, 2004)
11. W. Greiner, B. Muller, J. Rafelski, *Quantum Electrodynamics of Strong Fields: Introduction to Modern Relativistic Quantum Mechanics* (Springer, Berlin, 1985)

12. S.S. Bulanov et al., Phys. Rev. Lett. **104**, 220404 (2010)
13. A.H. Zewail, J. Phys. Chem. A. **104**, 5660 (2000)
14. A.D. Bandrauk, S. Barmaki, S. Chelkowski, G.L. Kamta, in "*Progress in Ultrafast Intense Laser Science*", vol III, Chap. 9, ed. By K. Yamanouchi et al. (Springer, N. Y., 2007)
15. S. Chelkowski, G. Yudin, A.D. Bandrauk, J. Phys. B. **39**, S409 (2006)
16. A.D. Bandrauk, J. Manz, M. Vrakking, Chem. Phys. **366**, 1 (2009)
17. Z. Chang, "*Fundamentals of Attosecond Optics*" (CRC Press, Taylor-Francis, USA, 2011)
18. A. Zavriyev, P.H. Bucksbaum, in "*Molecules in Laser Fields*", Chap. 2, ed. by A.D. Bandrauk (Marcel Dekker Inc., NY, 1994)
19. A.D. Bandrauk et al, in "*Molecules in Laser Fields*", Chap. 4, ed. by A.D. Bandrauk (Marcel Dekker Inc., NY 1994)
20. R.S. Mulliken, J. Chem. Phys. **7**, 20 (1939)
21. C. Cornaggia in "*Progress in Ultrafast Intense Laser Science*", vol VI, Chap. 1, ed. by K. Yamanouchi et al (Springer, Tokyo 2011)
22. T. Zuo, A.D. Bandrauk, Phys. Rev. A54, 2511 (1995); 54, 3254 (1996)
23. S. Chelkowski, A.D. Bandrauk, J. Phys. B **28**, L723 (1995)
24. T. Seideman, M.Y. Ivanov, P.B. Corkum, Phys. Rev. Lett. **75**, 2819 (1995)
25. N.B. Delone, V.P. Krainov, "*Multiphoton Processes in Atoms*" (Springer, Berlin, 1994)
26. M.J. DeWitt, R.J. Levis, J. Chem. Phys. **108**, 1, (1998)
27. T.F. Gallagher, Phys. Rev. Lett. **61**, 2304 (1988)
28. S. Chelkowski, J. Goudreault, A.D. Bandrauk, J. Mod. Opt. **52**, 411 (2005)
29. A.D. Bandrauk, S. Chelkowski, K.J. Yuan, Intl. Rev. Atom Molec. Phys. (2011)
30. G. Mourou, T. Tajima, Science **331**, 41 (2011)
31. U.W. Rathe, C.H. Keittel, M. Protopapas, P.L. Knight, J. Phys. B **30**, L531 (1997)
32. S. Selsto, E. Lindroth, J. Bengtsson, Phys. Rev. **A79**, 043418 (2009)
33. A. Maquet, R. Grobe, J. Mod. Opt. 49, 2001 (2002)
34. A.D. Bandrauk, H. Kono, in "*Advances in Multiphoton Processes*", vol 15, ed. by S.H. Lin (World Scientific, Singapore, 2000)
35. P. Dietrich, M.Y. Ivanov, F.A. Ilkov, P.B. Corkum, Phys. Rev. Lett. **77**, 4150 (1996)
36. K. Codling, L.J. Frasinski, P.A. Hatherly, J. Phys. B22, L321 (1989)
37. J.H. Posthumus et al., J. Phys. B. **28**, L 349 (1995)
38. J.H. Posthumus, Rep. Prog. Phys. **67**, 623 (2004)
39. H. Schröder, C.J. Uiterwaal, K.L. Kompa, Laser Phys. **10**, 749 (2000)
40. G.L. Kamta, A.D. Bandrauk, Phys. Rev. A **75**, 041401 (2007)
41. H.Z. Lu, A.D. Bandrauk, J. Molec. Str. (TheoChem) **547**, 97 (2001)
42. H. He et al, Phys. Rev. A. **84**,033418 (2011)
43. H. Sabzyan, M. Vaface, Phys. Rev. A. **71**, 063404 (2005)
44. M. Schmidt, D. Normand, C. Cornaggia, Phys. Rev. A. **50**, 5037 (1994); *ibid*, 53, 1958 (1996)
45. G.N. Gibson, M. Li, C. Guo, J. Neira, Phys. Rev. Lett. **79**, 2022 (1994)
46. K. Yamanouchi, Science 295, 1659 (2002)
47. G.L. Kamta, A.D. Bandrauk, Phys. Rev. Lett. 94, 203003 (2005); Phys. Rev. A 76, 053409 (2007)
48. X.B. Bian, A.D. Bandrauk, Phys. Rev. Lett. **105**, 093903 (2010)
49. Y.J. Jiu, X.M. Tong, N. Toshima, Phys. Rev. A. **81**, 013408 (2010)
50. A.D. Bandrauk, H.Z. Lu, Phys. Rev. A. **62**, 053406 (2000)
51. D. Pavicic, A. Kiess, T.W. Hänsch, H. Figger, Phys. Rev. Lett. **94**, 163002 (2005)
52. K. Sandig, H. Figger, T.W. Hänsch, Phys. Rev. Lett. **85**, 4876 (2000)
53. V. Serov et al., Phys. Rev. A. **72**, 033413 (2005)
54. E. Constant, H. Staplefeldt, P.B. Corkum, Phys. Rev. Lett. **76**, 4140 (1996)
55. J.N. Gibson, M. Li, C. Zuo, J. Neira, Phys. Rev. Lett. **79**, 2002 (1997)
56. A. Hishikawa, A. Iwamae, K. Yamanouchi, Phys. Rev. Lett. **83**, 1127 (1999)
57. C. Siedschlag, J.M. Rost, Phys. Rev. A 60, 2215 (1999); Phys Rev Lett. 89, 173401 (2002); Few-Body Systems, 31, 211 (2002)
58. I. Last, J. Jortner, Phys. Rev. A. **60**, 2215 (1999)

59. A.D. Bandrauk, G. Paramonov, AIP Conf. Proc. **1209**, 7 (2010)
60. Y. Sato, H. Kono, S. Koseki, Y. Fujimura, J. Am. Chem. Soc. **125**, 8019 (2003)
61. Y. Nagata et al., Phys. Rev. Lett. **71**, 3774 (1993)
62. L.N. Gaier, J. Mod. Opt. **52**, 1019 (2005)
63. A.D. Bandrauk, J. Ruel, Phys. Rev. A. **59**, 2153 (1999)
64. H. Yu, T. Zuo, A.D. Bandrauk, J. Phys. B. **31**, 1533 (1998)
65. I. Ben-Itzhak et al., Phys. Rev. A. **78**, 063419 (2008)
66. M. Ergler et al., Phys. Rev. Lett. **95**, 093001 (2005)
67. A.D. Bandrauk, S. Chelkowski, I. Kawata, Phys. Rev. A. **67**, 013407 (2003)
68. A. Staudte et al., Phys. Rev. Lett. **98**, 073003 (2007)
69. S. Chelkowski et al., Phys. Rev. A. **76**, 013405 (2007)
70. F. He, A. Becker, U. Thumm, Phys. Rev. Lett. **101**, 213002 (2008)
71. N. Takemoto, A. Becker, Phys. Rev. Lett. 105, 203004 (2010), Phys Rev A 84,023401(2011)
72. T. Zuo, A.D. Bandrauk, P. Corkum, Chem. Phys Lett. **259**, 313 (1996)
73. D. N. Fittinghoff, P.R. Bolton, B. Chang, R.C. Kulander, Phys. Rev. Lett. **69**, 2642
74. A.D. Bandrauk, S. Chelkowski, H. Yu, E. Constant, Phys. Rev. A. **56**, 2537 (1997)
75. P. Moreno, L. Playa, L. Roso, Phys. Rev. A. **55**, 1593 (1997)
76. A.D. Bandrauk, S. Barmaki, G.L. Kamta, Phys. Rev. Lett. **98**, 013001 (2007)
77. H. Yu, T. Zuo, A.D. Bandrauk, Phys. Rev. A 54, 3290 (1996); J Phys B 31, 1533 (1998)
78. I. Kawata et al., Phys. Rev. A 62, 031401 (2000); 66, 043403 (2002)
79. A. Saenz, Phys. Rev. A 61, 051402 (2000); 66, 063407 (2002)
80. I. Kawata, H. Kono, A.D. Bandrauk, Phys. Rev. A. **64**, 043411 (2001)
81. D. Tchitchekova, S.C. Chelkowski, A.D. Bandrauk, J. Phys. B. **44**, 065601 (2011)
82. S. Saugaut et al., Phys. Rev. Lett. **98**, 253003 (2007)
83. S. Sukiasyan et al., Phys. Rev. Lett. **102**, 223002 (2009)
84. E. Dehghanian, A.D. Bandrauk, G.L. Kamta, Phys. Rev. A. **81**, 061403 (2010)
85. A. Tong et al., Opt. Exp. **18**, 9064 (2010)
86. S.A. Trushin, W.E. Schmid, W. Fuss, J. Phys. B. **44**, 165602 (2011)
87. G.N. Gibson, M. Li, C. Guo, J.P. Nibarger, Phys. Rev. A. **58**, 4123 (1998)
88. I.V. Litvinyuk et al., Phys. Rev. Lett. **94**, 033003 (2005)
89. Y. Liu et al., Phys. Rev. Lett. **106**, 073004 (2011)
90. J.C. Slater, *Quantum Theory of Molecules and Solids*, vol I (McGraw-Hill, NY, 1963)
91. S. Chelkowski, A.D. Bandrauk, Phys. Rev. Lett. **101**, 153901 (2008)
92. E. Dehghanian, A.D. Bandrauk, G.L. Kamta, Phys. Rev. A **81**, 061403(R) (2010)
93. T.A. Green et al., J. Chem. Phys. **61**, 5156 (1974)
94. A. Etches, L.B. Madsen, J. Phys. B. **43**, 155602 (2005)
95. F. Légaré et al., Phys. Rev. A 72, 052717 (2005); A 71, 013415 (2005)
96. A. Hishikawa et al., Phys. Rev. Lett. **99**, 258302 (2007)
97. I. Bocharova et al., Phys. Rev. Lett. **107**, 063201 (2011)
98. A. Hishikawa, A. Iwamae, K. Yamanouchi, Phys. Rev. Lett. **83**, 1127 (1999)
99. J.P. Brichta, S.J. Walker, R. Helsten, J.H. Sanderson, J. Phys. B. **40**, 117 (2007)
100. E.P. Fowe, A.D. Bandrauk, Phys. Rev. A. **81**, 023411 (2010)
101. F. Légaré et al., Phys. Rev. Lett. **91**, 093002 (2003)
102. J.P. Brichta, A.N. Seaman, J.H. Sanderson, Comput. Phys. Commun. **180**, 197 (2009)
103. M. Okunishi et al., J. Chem. Phys. **127**, 064310 (2007)
104. M. Meckel et al., Science **320**, 1478 (2008)
105. Y. Chen, J. Chen, J. Liu, Phys. Rev. A. **74**, 063405 (2006)
106. Y. Huismans et al., Science **331**, 61 (2011)
107. M. Abel et al., Laser Photonics Rev. **5**, 352 (2011)

Chapter 3
Ultrafast Optical Gating by Molecular Alignment

Heping Zeng, Peifen Lu, Jia Liu, and Wenxue Li

Abstract Field-free alignment of gaseous molecules could function as an ultrafast polarization optical gating with periodic revivals originated from quantum wakes of the impulsively excited molecular wave packets. Recent experimental explorations have revealed some unique applications of ultrafast optical gating from pre-excited molecular rotational wave packets, such as molecular-alignment-based cross-correlation frequency-resolved optical gating (M-XFROG) for ultrashort pulse characterization, and molecular-alignment-based ultrafast optical imaging and optical buffer with revivable optical storage in molecular rotational wave packets. The M-XFROG technique employs the impulsive transient alignment of gaseous molecules as a gate function to characterize the ultrashort pulse and exhibits the advantage of no phase-matching constraint and applicability to pulses at any wavelength ranging from ultraviolet to far-infrared. Ultrashort pulse meaurements of ultraviolet pulse, supercontinuum pulse, optical parametric amplifier, and multi-colored pulses were experimentally performed by using the M-XFROG technique. Ultrafast optical imaging by periodic molecular alignment was also demonstrated, involving the optical image storage in the pre-excited molecular wakes followed by periodic readout and display. For diatomic molecules in air, both raised and intagliated monochromatic images were observed with periodic revivals of aligned molecules. Ultrafast time-encoded holographic-like imaging was realized to encode the phase information of a three-demensional object in the molecular revivals. The monochromatic images could be transformed into colorful optical imaging by using a spatially chirped supercontinuum laser pulse to chromatically encode the stored images with different colors at different delays with respect to the molecular alignment revivals.

H. Zeng (✉) · P. Lu · J. Liu · W. Li
State Key Laboratory of Precision Spectroscopy, East China Normal University, Shanghai 200062, China
e-mail: hpzeng@phy.ecnu.edu.cn

K. Yamanouchi et al. (eds.), *Progress in Ultrafast Intense Laser Science VIII*, Springer Series in Chemical Physics 103, DOI 10.1007/978-3-642-28726-8_3,
© Springer-Verlag Berlin Heidelberg 2012

3.1 Introduction

Field-free molecular alignment [1] with periodic revivals in response to quantum wakes of molecular wave packets is achieved through impulsive rotational Raman excitation by broadband ultrashort laser pulses. The periodic revivals originated from the rephasing of the pre-excited rotational wave packets provide not only field-free control on optical properties of the molecular gaseous medium but also robust methods to reveal molecular dynamics and structures for molecular physics and chemistry. So far, molecular alignment and orientation have sparked ever-growing research interest in various related fields, such as full-dimensional molecular manipulation [2], high-order harmonic generation [3], and molecular-orbital reconstruction [4]. Viewed from their intrinsic mechanisms, aligned and oriented molecular wave packets provide additional degrees of freedom to control the interaction and propagation of ultrashort laser pulses, which have already been demonstrated to bring about numerous intriguing applications in ultrashort pulse compression [5], alignment-induced focusing/defocusing and spatiotemporal phase modulation [6], central wavelength tuning of few-cycle ultrashort pulses [7], filamented propagation of intense femtosecond pulses [8–10], and filament interaction control. For instance, molecular alignment could facilitate intense femtosecond filament length elongation, observable enhancement of broadband supercontinuum generation, spatial regularization of multiple filaments, and so on.

After an advancing pump pulse excitation, the pre-aligned molecules evolve periodically and consequently lead to a spatial and temporal-dependent refractive index modulation. In particular, the ultrafast birefringence of the aligned molecules could act as a transient wave plate to change the polarization of a following probe pulse. Such a transient wave plate works with a period associated with the rephasing time of the pre-excited molecular rotational wave packets. For practical applications, this provides a new route to create an ultrafast polarization optical gating, which could be demonstrated by experiments using the so-called weak field polarization spectroscopic method. In this chapter, we review our recent experimental demonstrations on ultrafast polarization optical gating induced by field-free molecular alignment. The periodic revivals of the pre-excited rotational wave packets brought about revivable ultrafast molecular gating for the probe pulses, without any further requisite nonlinear couplings between the pump and probe pulses and accordingly, the ultrafast molecular gating and its periodic revivals could be intrinsically regarded as linear optical gating for the probe pulses. Such a linear ultrafast optical gating is anticipated to support some unique applications, such as ultrafast optical switching for weak pulses, ultrafast optical memories and revivable optical buffers, time-encoding for ultrafast holographic processing of optical information, and so on. Here, we review our experimental explorations on two alignment-relevant optical gating techniques: molecular-alignment-based cross-correlation frequency-resolved optical gating (M-XFROG) and ultrafast optical imaging by molecular alignment. Unlike the conventional frequency-resolved optical gating (FROG) geometries [11] based on nonlinear frequency mixing processes, no phase matching is required by

M-XFROG as molecular alignment functions in principle as a linear optical gating for the probe pulses, making M-XFROG a powerful technique for ultrashort pulse measurement. As ultrafast gated imaging in aligned gaseous molecules follows the periodic revivals of the pre-excited rotational wave packets, multiple functions such as the image storage in molecular wave packets, time-division phase encoding, periodic readout and holographic optical display are experimentally demonstrated.

3.2 Principle of Ultrafast Molecular Gating

3.2.1 Molecular Alignment Based Ultrafast Birefringence

A linear molecule exposed to a linearly polarized laser field can be approximately modelled as a linear rigid rotor. As described in Fig. 3.1, the electric field $\vec{E}$ of a linearly polarized laser pulse is along the z-axis, α is the molecular polarizability, // and $\perp$ represent the directions parallel and perpendicular to the molecular axis, and θ is the angle between the molecular axis and polarization direction of the pump pulse. According to the classical description, field-free molecular alignment arises as follows. An ultrashort laser pulse imparts an instantaneous "kick" on the molecules resulting in an induced dipole moment $\vec{P} = \vec{\alpha}\cdot\vec{E}$ which in return interacts with the laser field $\vec{E}$ Then, an angular-dependent potential energy $U(\theta)$ arises from the interaction between the induced dipole moment and the pump pulse field, which can be expressed as

$$U(\theta) = -\vec{P} \cdot \vec{E} = -\frac{1}{2}E^2[\alpha_\perp + \Delta\alpha \cos^2 \theta]. \tag{3.1}$$

Here, $\Delta\alpha = \alpha_{//} - \alpha_\perp$ is the polarizability difference between the components parallel and perpendicular to the molecular axis. Due to the nonzero polarizability difference along different directions, the pump field delivers a torque proportional to $-\mathrm{d}U/\mathrm{d}\theta$ to the molecules and forces them to rotate toward the polarization direction of the pump pulse. After the pump pulse excitation, the transient molecular alignment periodically revives in field-free condition.

3.2.2 Dynamics of Molecular Rotational Wave-Packets

Quantum mechanically, a "kick" of the pump pulse creates a rotational wave packet in the molecules. An ensemble of molecules which are initially randomly aligned will preferentially align along the polarization direction of the pump pulse soon after the "kick" However, along with the evolution of the rotational wave packet, the molecules rapidly lose their macroscopic alignment due to dephasing of the

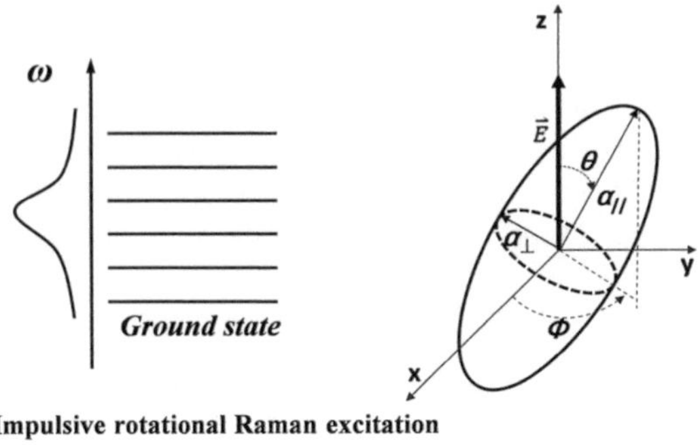

Fig. 3.1 Geometry of a prolate molecule in an electric field E and impulsive rotational excitation

rotational wave packet and subsequently keep in a state with an almost isotropic angular distribution until the next revival occurs.

Before the excitation of the pump pulse, each molecule is assumed to be in its ground state of angular momentum $|J, M\rangle$, where $J = 0, 1, 2, \ldots$, and $M = -J, -(J-1), \ldots, J-1, J$ account for the orbital momentum and the corresponding projection onto the symmetry axis of the molecule.

As shown in Fig. 3.1, when a molecule is off-resonant excited by a linearly polarized laser pulse, the interaction of the laser pulse with the linear molecule can be described by the effective Hamiltonian

$$H = H_0 - \frac{1}{4}\Delta\alpha E^2 \cos^2\theta, \tag{3.2}$$

with the field-free Hamiltonian

$$H_0 = B_0 J(J+1) + D_0 J^2 (J+1)^2, \tag{3.3}$$

where B_0 and D_0 are the rotational constant and centrifugal distortion constant, respectively.

The rotational wave function $\Psi(t)$ of the impulsively excited molecules could be derived from the Schrödinger equation [12]

$$\mathrm{i}\hbar\frac{\partial\Psi(t)}{\partial t} = H(t)\Psi(t). \tag{3.4}$$

The molecular rotation state $\Psi(t)$ can be expanded in the eigenstates as

$$\Psi(t) = \sum\nolimits_{JM} C_{JM}|J, M\rangle. \tag{3.5}$$

During the excitation of the laser pulse, the molecular rotational states could be calculated by numerically solving the time-dependent Schrödinger equation with the Cranck–Nicholson method.

After the impulsive Raman excitation, the population of each rotational state continues to evolve in the fieldfree Hamiltonian H_0 as

$$\Psi(t) = \sum_{JM} C_{JM}\, \mathrm{e}^{-\mathrm{i}E_J t} |J, M\rangle, \tag{3.6}$$

Where $E_J = B_0 J(J+1) + D_0 J^2(J+1)^2$ represents the eigenenergy of the field-free Hamiltonian. With negligible rovibrational coupling (i.e. $D_0 \sim 0$), the field-free energy reads as $E_J = B_0 J(J+1)$. The molecular rotational states evolve periodically as

$$\Psi(t = \pi/B_0) = \sum_{JM} C_{JM}\, \mathrm{e}^{-\mathrm{i}B_0 J(J+1)(\pi/B_0)} |J, M\rangle = \sum_{JM} C_{JM} |J, M\rangle = \Psi(0). \tag{3.7}$$

This means the field-free evolution of any rotational wave packet exhibits a periodicity $T_{\mathrm{rev}} = \pi/B_0$, i.e., the rotational wavepacket exactly reproduces itself at an integer multiples of the revival time, which accounts for the periodic revivals of the molecular alignment after the extinction of the pump pulse. Note that the revival period of the molecular alignment is determined by the rovibrational molecular constant B_0, independent upon the impulsive excitation pulses or initial thermal population distribution or impulsively excited state coherence. Different molecules thus exhibit their intrinsic revivals, while the time-dependent revival profiles are closely related with the pump pulse intensity and duration and initial thermal population on the rovibrational states. Moreover, as is well known, at rational fractions of the revival time, fractional rotational revivals of the rotational wave packets could also be observed [13]. The rational revivals typically occur at quarter, half, three-quarters of the corresponding molecular revival

3.2.3 *The Molecular Alignment Matrices*

The degree of molecular alignment could be well characterized by the quantity $\langle\langle \cos^2\theta \rangle\rangle$, which is calculated by a double averaging procedure. First the Schrödinger equation is solved for each initial molecular rotational state

$$|\Psi(t = 0)\rangle = |J_0 M_0\rangle, \tag{3.8}$$

and the time-dependent degree of molecular alignment is obtained from the population $C_{JM}(t)$ of each rotational state as

$$\langle \cos^2\theta \rangle_{J_0 M_0} = \sum_{J'M',JM} C^*_{J'M'} C_{JM} |J'M'\rangle \cos^2\theta |JM\rangle. \tag{3.9}$$

Then consider that the population of initial states follows the temperature-dependent Boltzmann distribution

$$\rho(t = 0) = \sum P_{J_0} |J_0 M_0\rangle\langle J_0 M_0|, \tag{3.10}$$

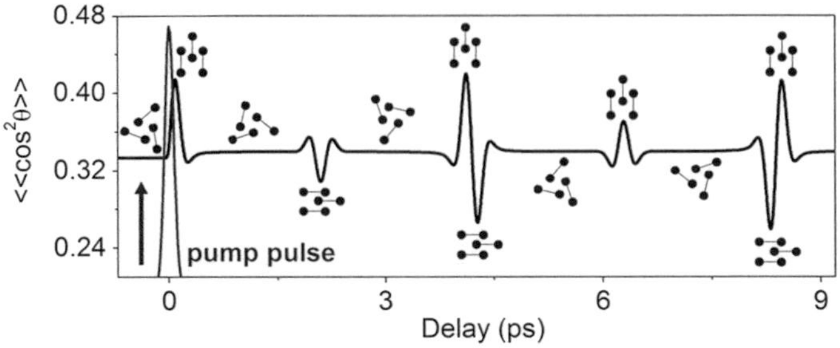

Fig. 3.2 Typical calculated molecular alignment signal of N_2

so the degree of molecular alignment during the field-free evolution is given by further average of $\langle \cos^2 \theta \rangle$

$$\langle\langle \cos^2 \theta \rangle\rangle = \frac{\sum_{J_0} P_{J_0} \sum\limits_{M_0=-J_0}^{J_0} \langle \cos^2 \theta \rangle_{J_0 M_0}}{\sum_{J_0} P_{J_0}}, \tag{3.11}$$

where P_{J_0} is the population probability of the rotational state $J0$.

Figure 3.2 shows a calculated evolution of impulsive molecular alignment of N_2. It can be seen obviously that the revivals of transient molecular alignment occur under the field-free condition after the pump pulse. The full revival period T_{rev} is 8.38 ps for N_2 and hence the alignment signal repeats every 8.38 ps. For randomly oriented molecules, the average term $\langle\langle \cos^2 \theta \rangle\rangle$ is equal to 1/3. The molecular orientation tends to be parallel to the field polarization when $\langle\langle \cos^2 \theta \rangle\rangle$ is greater than 1/3, whereas smaller values indicate perpendicular orientation.

3.2.4 Weak Field Polarization Spectroscopic Technique

There are two main techniques widely used in the experimental measurements of field-free molecular alignment. One is Coulomb explosion imaging [14, 15] and the other is weak field polarization spectroscopic method [16, 17]. Throughout this chapter, the alignment matrices are measured with the weak field polarization spectroscopic method, which reveals polarization variation of a weak probe pulse passing through prealigned molecules The intensity of the probe pulse is so weak that it produces little influence on the prealigned molecules.

Figure 3.3 shows the typical experimental setup used in the weak field polarization spectroscopic measurements of molecular alignment. Femtosecond laser pulses from a Ti:sapphire laser system are split into two beams: a strong pump beam for aligning gaseous molecules and a weak probe beam for probing the periodic revivals of molecular alignment after the pump beam. The polarization of the pump pulse is set to be 45° with respect to that of the probe pulse. The two beams are focused

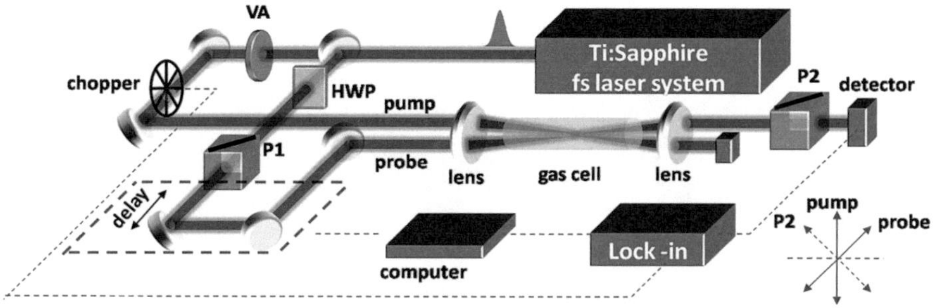

Fig. 3.3 Schematic of the experimental setup for molecular alignment measurement by using weak field polarization technique. *VA*: variable attenuator; *HWP*: half-wave plate; *P1 and P2*: polarizer

on the same point in the gas cell filled with gaseous molecules. The probe pulse experiences polarization modulation when passing through the aligned molecules due to the birefringence of the prealigned molecules. There are two polarizers (P1 and P2) with crossed transmission polarization laid before and after the gas cell in the probe beam. Hence, there exists no probe light transmission though the polarizer (P2) when the molecules are not aligned, whereas, a portion of the probe pulse can be transmitted as a signal pulse due to the polarization modulation induced by the aligned molecules. The intensity of the signal is determined by the degree of the molecular alignment and the evolution of the signal reflects the field-free evolution of the wave packets of the aligned molecules. Note that the measured signal is proportional to $(\langle\langle\cos^2\theta\rangle\rangle - 1/3)^2$ rather than $\langle\langle\cos^2\theta\rangle\rangle$

3.2.5 Alignment-Induced Polarization Optical Gating

In this section, we will demonstrate that femtosecond laserinduced molecular alignment with periodic revivals can function as a revivable ultrafast polarization optical gating owing to the birefringence of the aligned molecules.

3.2.5.1 Alignment-Induced Change of the Refractive Index

Field-free molecular alignment with periodic revivals gives rise to a variation of the refractive index along the polarization direction, which can be expressed as [18]

$$\Delta n_R(r,t) = 0.5(\rho\Delta\alpha/n_0)(\langle\langle\cos^2\theta\rangle\rangle - 1/3), \tag{3.12}$$

where t is the time delay after the excitation pulse ρ is the density of molecules, and n_0 is the linear refractive index. Hence, the refractive index varies with the evolution of the molecular alignment. When $\langle\langle\cos^2\theta\rangle\rangle > 1/3$, the molecules are

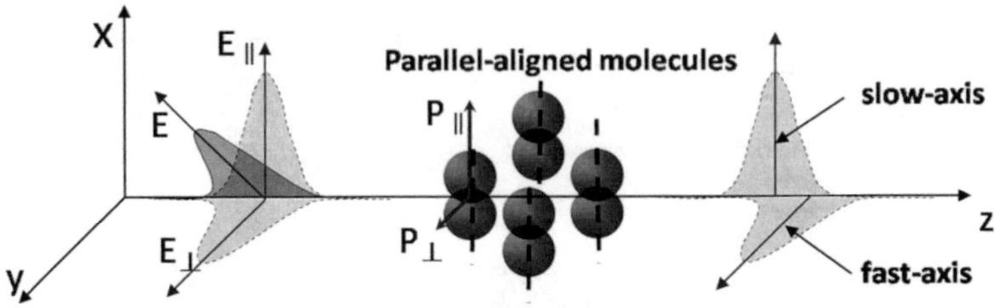

Fig. 3.4 Schematic of birefringence induced by molecular alignment

preferentially aligned along the pump polarization direction, accompanied with an increased refractive index along the polarization axis. When, the molecules are preferentially aligned in a plane perpendicular to the pump polarization direction, resulting in a decreased refractive index According to (3.12), the time-dependent refractive index gives rise to ultrafast transient birefringence, linearly dependent on the molecular alignment matrix. The ultrafast birefringence exactly follows the molecular revivals.

As depicted in Fig. 3.4, when a probe pulse propagates through the prealigned molecules, the crossed components parallel and perpendicular to the alignment direction undergo different refractive indices and different phase velocities, leading to a phase delay between the two components of the pulse. Thus the polarization state of the probe pulse is changed. Therefore, the prealigned molecules function analogously to a transient wave plate that can be used as an alignment-induced polarization gating.

3.2.5.2 Alignment-Induced Spatiotemporal Phase Modulation

According to (3.12), the transient molecular alignment spatially and temporally modifies the refractive index of the molecular gas, leading to a phase shift imposed on the probe pulse. Essentially, it can be considered that the pump pulse, mediated by the wake of the molecular alignment, induces a spatiotemporal cross-phase modulation of the probe pulse.

After the impulsive excitation driven by the pump pulse, the transient molecular alignment revives periodically in a field-free condition. As a consequence, a properly matched probe pulse undergoes an additional nonlinear phase shift φ_m ascribed to the revivals of molecular alignment, corresponding to a spectral phase modulation given by [6]

$$\delta\omega(t) = -\partial\varphi_m/\partial t \sim -\partial\delta n(t)/\partial t. \tag{3.13}$$

Moreover, each revival of the transient molecular alignment can be considered as an ultrashort optical gating. Owing to the recurrences of molecular alignment, the alignment-induced polarization optical gating has a periodicity with the same period as the revivals of molecular alignment.

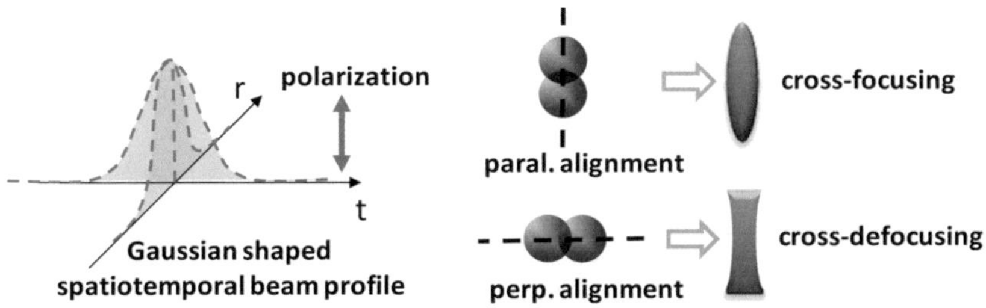

Fig. 3.5 Schematic representation of alignment-induced spatial cross (de)focusing

For a moderate intensity (an intensity below saturation of alignment), the degree of the molecular alignment ($\langle\langle\cos^2\theta\rangle\rangle - 1/3$) almost depends on the pump intensity linearly Thus, the refractive index profile induced by molecular alignment follows the spatial profile of the pump pulse. Consider a pump pulse with a Gaussian-shaped spatiotemporal beam profile as shown in Fig. 3.5, the degree of molecular alignment in the central part of the pump beam is larger than that in its periphery since the intensity is higher in the center, resulting in a deeper modulation of the refractive index of the corresponding molecular gas. So when the orientation of the molecules is parallel to the polarization of the pump pulse, the refractive index increases more in the center than in the periphery, equivalent to a convex lens. On the contrary, when orientation of the molecules is perpendicular to the polarization of the pump pulse, the refractive index decreases more in the center than in the periphery, equivalent to a concave lens. Correspondingly, a probe pulse properly matching the parallel or perpendicular orientation of the molecules during the revival will experience an additional cross focusing or defocusing effect. Therefore, the spatial distribution of aligned molecules causes a refractive index gradient that can be considered as an all-optical-controlled lens and interestingly, it is spatiotemporally modulated in a field-free way around the molecular revivals.

To conclude, we demonstrate that transient molecular alignment could function as a revivable ultrafast polarization optical gating based on the birefringence of the aligned molecules, providing a new route to ultrafast optical gating. In the following sections, two applications of the molecular alignment optical gating will be presented.

3.3 Molecular-Alignment-Based Cross-Correlation Frequency-Resolved Optical Gating

3.3.1 Frequency-Resolved Optical Gating

Consider an ultrashort pulse with a time-dependent component expressed as

$$E(t) = \mathrm{Re}\left\{\sqrt{I(t)}\,\exp(\mathrm{i}\omega_0 t - \mathrm{i}\varphi(t))\right\}, \tag{3.14}$$

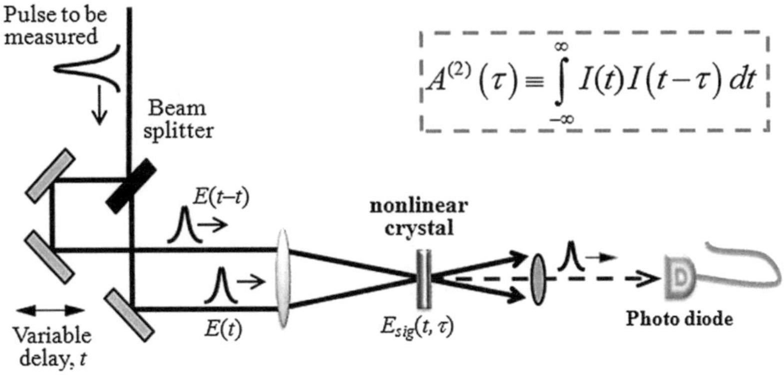

Fig. 3.6 Schematic of the intensity autocorrelator, where pulse duration could be measured from the time-correlated intensity profiles, while no pulse phase information could be revealed and no direct measurement of electric fields could be reached

where ω_0 is the carrier frequency and $I(t)$ and $\varphi(t)$ are the time-dependent intensity and phase of the pulse. So full characterization of ultrashort pulses needs detailed information of the intensity and phase of the pulse.

Autocorrelator is commonly used in time-domain characterization of a pulse, which involves overlapping two split pulses with variable delays in an instantaneously responding nonlinear optical crystal. The typical autocorrelator employs a setup as shown in Fig. 3.6, where the pulse duration could be measured from the time-correlation of the intensity profile of the pulse to be measured.

In a typical intensity autocorrelator based on second harmonic generation, by measuring the signal with the time delay τ, intensity correlation can be obtained from

$$A(t) = \int_{-\infty}^{+\infty} I(t)I(t-\tau)\,\mathrm{d}t. \tag{3.15}$$

The auto-correlation signal could reveal the pulse duration with some preknowledge of the pulse profile. While $A(t)$ contains only the pulse intensity profiles, no phase coherence could be deduced. Accordingly, the standard autocorrelators merely provide pulse duration measurement from the time-correlated pulse intensity profiles rather than direct measure of the electric fields.

Compared with the autocorrelator, the FROG technique solves the phase-losing problem and provides a straightforward and complete temporal measurement of an ulrashort pusle. To date, the FROG technique has been one of the most popular and robust methods for full characterization of ultrashort laser pulse.

As shown in Fig. 3.7, the FROG technique is quite simple to implement in experiments. The experimental setup is almost the same with the intensity autocorrelator. The main difference is the signal collection. FROG collects a spectrogram by spectrally resolving the gated pulse versus the time delay, as illustrated in Fig. 3.8a. The intrinsic mechanism of FROG thus involves optically gating a target

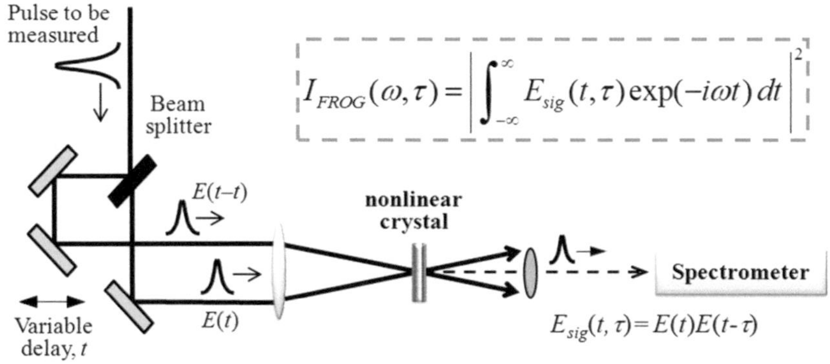

Fig. 3.7 A schematic of FROG based on second harmonic generation, which involves gating the pulse with a variably delayed replica of itself in a nonlinear optical medium and spectrally resolving the gated pulse versus delay

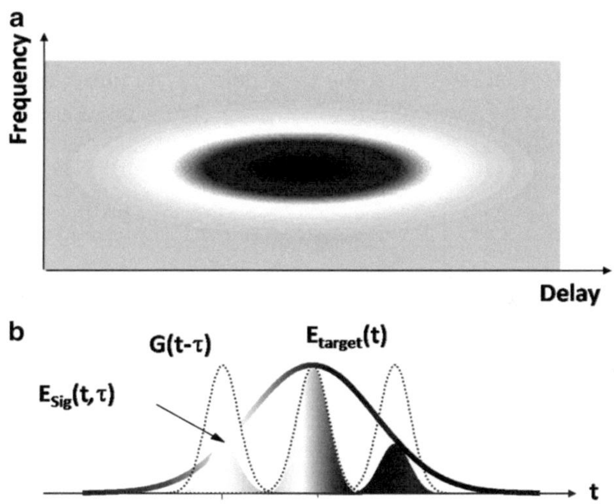

Fig. 3.8 (**a**) A typical FROG trace and (**b**) Schematic presentation of the gate

pulse (i.e., the pulse to be measured) with a variable time-delayed gate function in a nonlinear optical medium, as schematically shown in Fig. 3.8b. After the spectrogram measurement, FROG relies on a mathematical retrieval of the electric field amplitude and phase for the pulse to be measured from the spectrogram by a retrieval algorithm.

As schematically presented in Fig. 3.8b, the signal pulse (the gated pulse) in the time domain can be given by

$$E_{\text{Sig}}(t,\tau) = E_{\text{Target}}(t)G(t-\tau) \tag{3.16}$$

Table 3.1 Various FROG schemes with the corresponding optical nonlinearities involved in the measurement

Various FROG geometries

FROG	Geometries	$E_{sig}(t,\tau)$	Nonlinearity
PG FROG	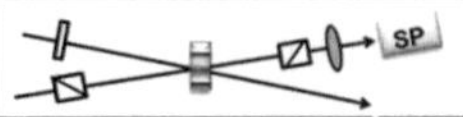	$E(t)^2\|E(t-\tau)\|^2$	$\chi^{(3)}$
SD FROG		$E(t)E^*(t-\tau)^2$	$\chi^{(3)}$
SHG FROG		$E(t)E(t-\tau)$	$\chi^{(2)}$
THG FROG	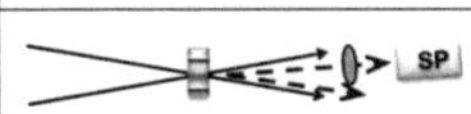	$E(t)^2E(t-\tau)$	$\chi^{(3)}$
XFROG	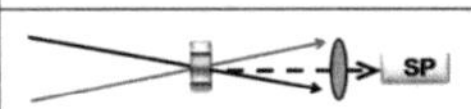	$E(t)G(t-\tau)$	$\chi^{(3)}$

where $E_{\text{Target}}(t)$ is the electric field of the target pulse to be measured and $G(t-\tau)$ is the gate function with respect to the time delay τ. Figure 3.8a shows a typical FROG trace obtained by spectrally resolving the signal pulse versus delay, which can be generally written as

$$I_{\text{FROG}}(\omega,\tau) = \left| \int_{-\infty}^{\infty} E_{\text{Target}}(t)G(t-\tau)\exp(-\text{i}\omega t)\,\text{d}t \right|^2. \quad (3.17)$$

If $G(t)$ is a function of $E_{\text{Target}}(t)$ itself, it is called FROG technique; if not, it is termed as cross-correlation frequency-resolved optical gating (XFROG), which is usually preferable to gate the target pulse with a known or different pulse. As summarized in Table 3.1, a variety of geometries have been developed for FROG measurements for different specific applications, such as the polarization gating [19], self-diffraction [20], second harmonic (SH) generation [21], third harmonic (TH) generation [22], and transient grating [23] geometries. Table 3.1 also lists various FROG geometries and the related optical nonlinearities involved for the electric field characterization.

Generally speaking, a nonlinear optical crystal is required in nonlinear frequency-mixing of gate and target pulses. Such nonlinear interaction unavoidably sets some limitations for ultrashort pulse measurement, such as attainable phase-matched spectral bandwidth, group-velocity mismatch, frequency conversion efficiency, transparent spectral range of the crystal, and material-dispersion induced broadening of the target pulses. The standard FROG technique is limited by some practical issues as follows:

- Measurement sensitivity is limited due to the requisite optical nonlinearity in coupling the target and gate pulses and sufficiently high signal to noise ratio for spectrogram signal required by FROG retrieval algorithm. Nonlinear interaction usually requires sufficiently strong input pulses, and it becomes quite difficult to get FROG signals for too weak target pulses. The FROG retrieval algorithm needs a sufficiently high signal to noise ratio for the FROG spectrogram, otherwise the retrieved electric field might deviate from the target pulse and yield a meaningless FROG retrieval.

- Phase-matching spectral band limits the maximum spectral bands that can be measured by FROG. This becomes extremely serious when few-cycle femtosecond pulses or ultrabroadband supercontinuum pulses are concerned. The FROG implementation actually needs critic optical designs to extend the measurable spectral bands, such as ultrathin nonlinear crystals, achromatic phase-matching conditions, and so on.
- Material dispersion sets a temporal limit for FROG as a large dispersion may distort the temporal profile for ultrashort pulses. The temporal resolution of a standard FROG is restricted by the dispersion and thus pulse profile distortion from the nonlinear optical crystals where the target pulses have to interact with the gate pulses.
- Nonlinear optical process limits the specific working wavelength range of the standard FROG. This limit mainly comes from the fact that the phase matching condition is only fulfilled at a specific wavelength range of the target pulses. As the central wavelength of the target pulses is tuned beyond the working range, FROG needs to change the nonlinear optical crystals with corresponding changes of the whole optical setup. It is still a challenge to make a single FROG to cover a broadband working range for central wavelength tunable target pulses.

All the above limitation actually comes from the requisite nonlinear coupling between target and gate pulses. If a process works with a linear process instead of a nonlinear process, the target pulse could thus be gated with an improved sensitivity, and accordingly, no phase matching is required as optical gating works without nonlinear processes. To extend its applicable spectral range, we need an optical gating process to work efficiently in gating target pulses at arbitrary wavelengths from UV to far infrared. In addition, negligible dispersion should be accompanied with the optical gating process, which should be better implemented in gas rather than optical crystals. The key question of practical importance is whether it is possible to realize a FROG based on a linear process applicable from UV to far infrared. As well demonstrate in what follows, impulsive alignment of gaseous molecules can function as linear optical gating for ultra-broadband pulses or tunable pulses from UV to far infrared.

3.3.2 XFROG by Molecular Alignment Optical Gating

Due to the quantum beatings of the rotational wave packets of the impulsively pre-excited molecules, there are periodic revivals of molecular alignment parallel or perpendicular to the polarization of the pump pulse after the excitation. As we have mentioned, for a probe pulse precisely matched in time domain to any revival of the transient molecular alignment, it can feel the birefringence effect of the aligned molecular and consequently its polarization state would be changed. This provides a route to develop a new XFROG geometry for ultrashort pulse measurement by using the molecularalignmentinduced optical gating as a gate function. As compared with the conventional FROG geometries, the molecular-alignment-gated XFROG utilizes the transient molecular alignment as a gate and makes no use of any nonlinear crystal for nonlinear frequency conversion. M-XFROG exhibits quite a lot of distinguished advantages that make it work as a robust technique for ultrashort laser pulse measurement. Note that the gating process of the M-XFROG can be carried out in a gaseous medium such as air, which makes the implementation of M-XFROG quite convenient. In particular, the gaseous molecules induce quite low dispersion to the target pulse as compared to the nonlinear optical crystals. This facilitates direct electric field characterization of ultrabroadband pulses or few-cycle pulses with negligible distortion from the measurement processes. Although the molecular alignment is intrinsically a nonlinear process, the gating process on the target pulse is indeed linear and adds no phase information to the gated slice of $E_{\mathrm{Target}}(t)$, and accordingly, no phase-matching is required in molecular-alignment-based linear gating. Ultrashort pulses at any central wavelengths in the transparent range of the gases can in principle be measured by using M-XFROG. The linear-gating process in principle sets almost no intensity constraint on the target pulse as the molecular gating only depends on the pre-excited molecules driven by the pump pulses and its revivals occurs as long as the target pulse is properly matched in the revival periods, and thus ultrashort pulses of quite weak intensity can be measured.

Figure 3.9 presents the schematic of the M-XFROG which uses transient molecular alignment as an optical gating based on the weak field polarization technique. A target pulse successively passes through two crossed polarizers placed before and after the molecular gas (e.g., air) so that no signal pulse transmits in the case of random alignment. A linearly polarized pump pulse is set to be 45° with respect to that of the target pulse. These two pulses are spatially overlapped where the optical gating process occurs. After the pump pulse excitation, the time-dependent gate is created essentially on the basis the transient birefringence of the aligned molecules. When the target pulse is properly tuned to match with a revival of the transient molecular alignment, aligned molecules act as a transient wave plate [24] and the gating process takes place linearly by means of rotating the polarization of the target pulse which is analogous to the cross-correlation polarization-gating FROG in bulk media. Consequently, a portion of the target pulse transmits through the polarizer as a signal pulse and is then sent to be spectrally resolved versus delay by a spectrometer to form an M-XFROG trace.

In principle, any revival of the transient molecular alignment can be used as a gate function. We used the molecular alignment around zero time delay as a gate since the signal of molecular alignment reaches its maximum soon after the pump pulse excitation. The width of the molecularalignmentinduced gate is related with the bandwidth of the excitation pulse, molecular gas temperature and molecular species.

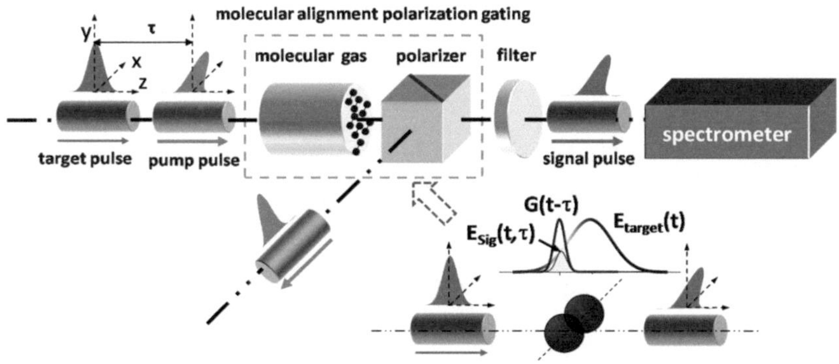

Fig. 3.9 Schematic of molecular-alignment-based frequency-resolved optical gating (M-XFROG)

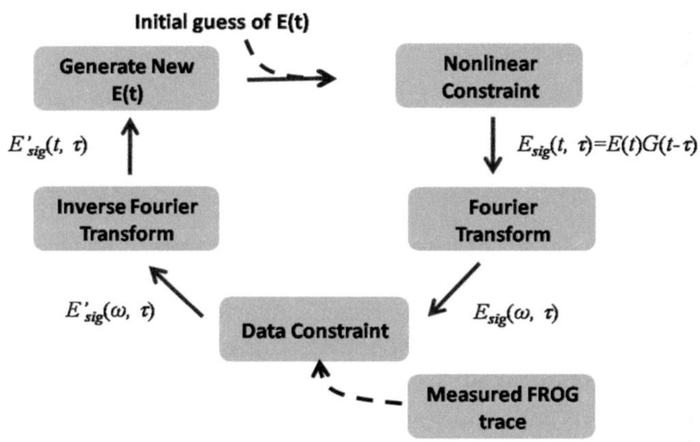

Fig. 3.10 Schematic of the generalized projection iterative algorithm

The XFROG spectrogram obtained from the experiment provides immediate qualitative information of the target pulse such as the central wavelength, spectral bandwidth, and chirp of the pulse. In order to further reveal the details of the unknown target pulse $E_{\mathrm{Target}}(t)$ from the measured M-XFROG trace, an iterative Fourier-transform algorithm with a generalized projection method is implemented [25]. As shown in Fig. 3.10, the retrieval procedure runs as follows. For the first projection, the trace signal $E_{\mathrm{Sig}}(t,\tau)$ is constructed as a convolution of a guessed electric field $E_{\mathrm{Target\_Guess}}(t)$ and the input gate function $G(t)$. And then, $E_{\mathrm{Sig}}(t,\tau)$ is Fourier-transformed with respect to t to obtain the $\tilde{E}_{\mathrm{Sig}}(\omega,\tau)$ since the XFROG trace is given as a function of ω and τ. Next, $\tilde{E}_{\mathrm{Sig}}(\omega,\tau)$ is fit to the XFOG trace by replacing its amplitude with the square root of the measured M-XFROG trace $I_{\mathrm{XFROG}}(\omega,\tau)$, keeping the phase unchanged. After that, the modified field $\tilde{E}_{\mathrm{Sig}}(\omega,\tau)$ is transformed back to time domain by inverse Fourier transform to obtain a new trace signal $E'_{\mathrm{Sig}}(t,\tau)$ for the next projection.

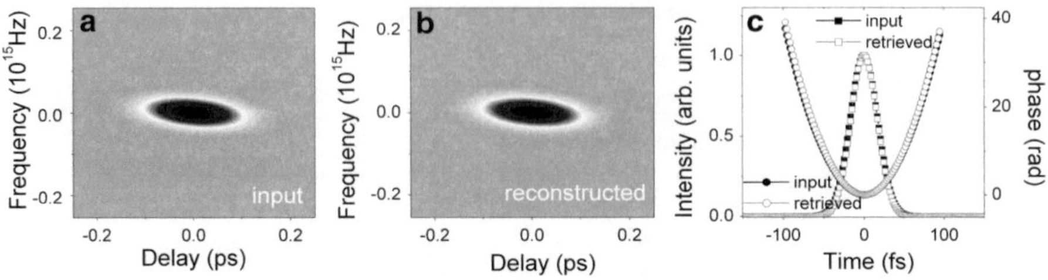

Fig. 3.11 (**a**) Numerically constructed XFROG trace from an assumed negatively chirped Gaussian pulse of 40 fs (FWHM) and a molecular alignment gate of 200 fs (FWHM). (**b**) Retrieved XFROG trace. (**c**) Comparisons of temporal intensity and phase between the input and retrieved pulses

In order to find a proper new $E'_{\mathrm{Target}}(t)$ for the next generalized projection, a distance metric is defined as

$$Z = \sum_{i,j=1}^{N} \left| E'_{\mathrm{Sig}}(t_i, \tau_j) - E'_{\mathrm{Target}}(t_i) G(t_i - \tau_j) \right|^2. \tag{3.18}$$

The iterative process is repeated to minimize Z with respect to $E'_{\mathrm{Target}}(t)$ until the XFROG trace error reaches an acceptable minimum [11].

To evaluate the validity of our retrieval algorithm for the XFROG trace, a FROG trace was numerically constructed according to an assumed pulse, and then a retrieved pulse was obtained from the FROG trace by our retrieval algorithm. The input and retrieved pulses shown in Fig. 3.11 clearly confirms the capacity and validity of the retrieval algorithm for ultrashort laser pulse measurement.

3.3.3 Experimental Results by M-XFROG

Unlike the conventional FROG geometries, no nonlinear crystals are required by the M-XFROG so that it shows unique features for ultrashort pulse characterization. A series of experiments were carried out to measure the UV, tunable OPA and supercontinuum (SC) pulses by using the M-XFROG, indicating that the molecular-alignment-gated XFROG is a powerful technique for the measurement of ultrashort laser pulses at any wavelength ranging from UV to far infrared.

Figure 3.12 presents experimental results of the measurements for weak UV pulses around 267 nm (third harmonic, TH) and 200 nm (fourth harmonic, FH). It can be seen from the results that ultrashort extreme UV pulse can be fully characterized by M-XFROG [26, 27]

Furthermore, experimental measurements of optical parametric amplifier (OPA) pulses at 400, 550, and 700 nm were shown in Fig. 3.13, which demonstrate that

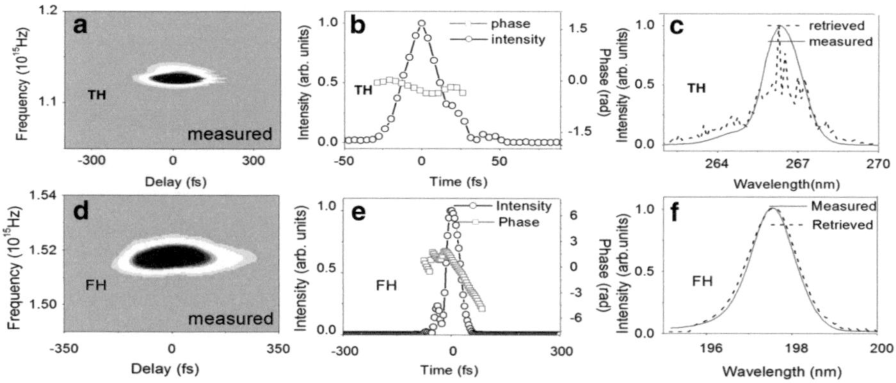

Fig. 3.12 Measurements of UV pulses by M-XFROG. (**a**) and (**d**) are measured XFROG traces. (**b**) and (**e**) are the retrieved temporal intensity and phase.(**c**) and (**f**) are the measured and retrieved spectra of the target pulse. *TH*: third harmonic; *FH*: fourth harmonic

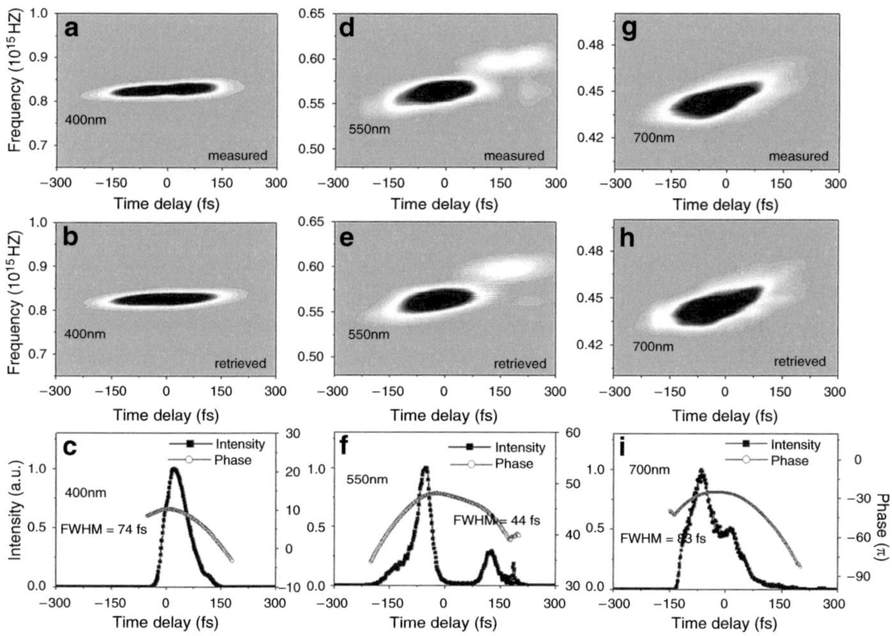

Fig. 3.13 M-XFROG for the measurements of OPA output at 400 nm, 550 nm and 700 nm. (**a**), (**d**) and (**g**) are the measured M-XFROG traces. (**b**), (**e**), and (**h**) are the retrieved M-XFROG traces. (**c**), (**f**), and (**i**) are the retrieved temporal intensity and phase of the target pulses

the molecular-alignment-induced optical gating is applicable for arbitrary central wavelengths from UV to infrared.

Full diagnose of a complex weak ultrabroadband supercontinuum pulse is still a tough task up to now because of the broadband phase-matching restriction. Recently,

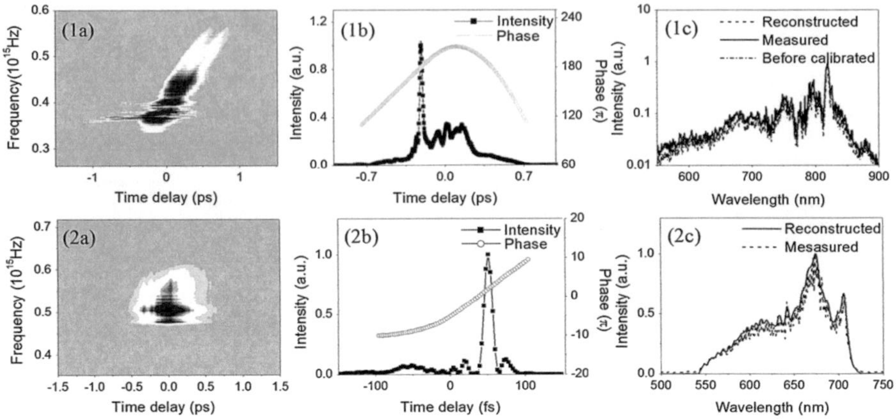

Fig. 3.14 M-XFROG for the measurement of SC pulses. (1a), (1b), and (1c) are experimental and retrieved results for the positively chirped SC pulse, respectively, corresponding to the measured M-XFROG trace, the retrieved temporal intensity and phase, and the measured and retrieved spectra. (2a), (2b), and (2c) are experimental and retrieved results for the compressed SC pulse by using two pairs of chirped mirrors, respectively, corresponding to the measured M-XFROG trace, the retrieved temporal intensity and phase, and the measured and retrieved spectra

a method has been proposed to achieve broadband phase matching by angle-dithering a thin nonlinear crystal [28]. For a direct comparison, M-XFROG exhibits obvious convenience in measuring supercontinuum pulses since no phase matching is required [29]. Figure 3.14 presents our experimental and retrieved results for a positively chirped supercontinuum pulse and a compressed supercontinuum pulse by using two pairs of chirped mirrors. It was clearly demonstrated that the M-XFROG can be not only used to characterize the complex long pulse but also versatile for diagnosis of ultrashort laser pulses in various spectral regions.

Interestingly, M-XFROG was further demonstrated to be available for simultaneous measurement of pulses consisting of different wavelengths at one spectrogram and additionally figure out the time delay between the pulses [27]. It should be pointed out that these different pulses should be controlled to have the same polarization in the practical measurements. In our experiment, a target pulse composed of 267 nm (third harmonic, TH) and 800 nm (fundamental wave, FW) pulses was measured and the results are shown in Fig. 3.15

3.3.4 *Discussions on M-XFROG Applicability*

During the practical applications, some important issues must be pointed out and discussed. Some critic concerns on the M-XFROG applicability and influence from nonlinear effects other than molecular alignment are summarized as follows:

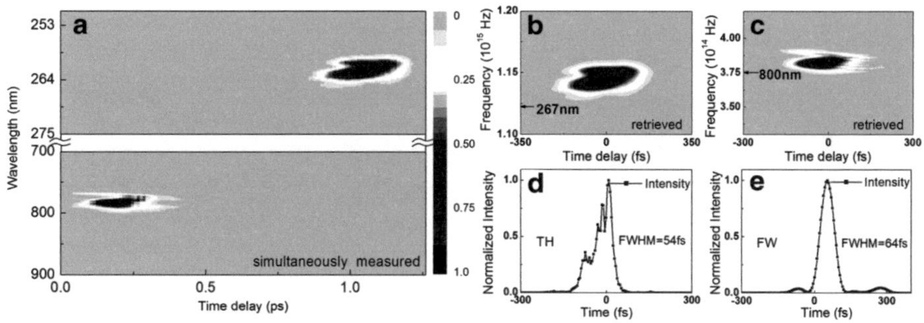

Fig. 3.15 M-XFROG for the simultaneous measurement of pulses at 267 nm (TH) and 800 nm (FW). (**a**) The measured M-XFROG trace containing SH and FW pulses. (**b**) and (**d**) are retrieved M-XFROG trace and temporal intensity of the SH pulse. (**c**) and (**e**) are retrieved M-XFROG trace and temporal intensity of the FW pulse

- The reliability of the M-XFROG was confirmed by a good agreement among the results obtained with M-XFROG, auto-correlation (AC) measurement, and spectral phase interferometry for direct electric field reconstruction (SPIDER) [27]. Even the ultrashort pulse with duration smaller than the gate width can be characterized by the M-XFROG, it was experimentally demonstrated that the M-XFROG could be used to measure few-cycle pulse of $\sim$ 12.8 fs [29].
- The contribution of the instantaneous Kerr effect during the pulse duration to the gate function was as small as compared with the molecular alignment effect [30] and it was also experimentally confirmed by our experiments [31, 32].
- The wavelength response, acquired by comparing the directly measured spectrum of the supercontinuum pulse and the time marginal of the M-XFROG trace (Fig. 3.16b), has been investigated in our experiment for wavelength calibration [29], as shown in Fig. 3.16c.
- As shown in the inset of Fig. 3.16d, the modulation efficiency, defined as a ratio of the pulse energy of the portion of the target pulse with polarization modulation to that of the whole input target pulse, presents a nearly linear response to the incident power. Meanwhile, the modulation efficiency exhibits a nonlinear increase with the increase of the excitation pulse intensity, mainly due to the saturation of the molecular alignment.
- The molecular- alignment-induced gate is insensitive to the slight fluctuation of the excitation pulse. As shown in Fig. 3.16e, we assumed three different excitation pulses with the same power but different pulse duration and peak intensity. The corresponding induced molecular alignment gate is shown in Fig. 3.16f; a clearly similar gate strongly supports the claim of the insensitivity of the molecular-alignment-induced gate and thus enables a robust measurement of the ultrashort target pulse with no requisite strict control on the pump excitation.
- There should be a limit of the lowest intensity of the pulse that can be resolved in the practical experiment condition. The sensitivity is estimated to be $\sim$ nJ by considering that the spectrometer sensitivity ($\sim$100 photons/count),

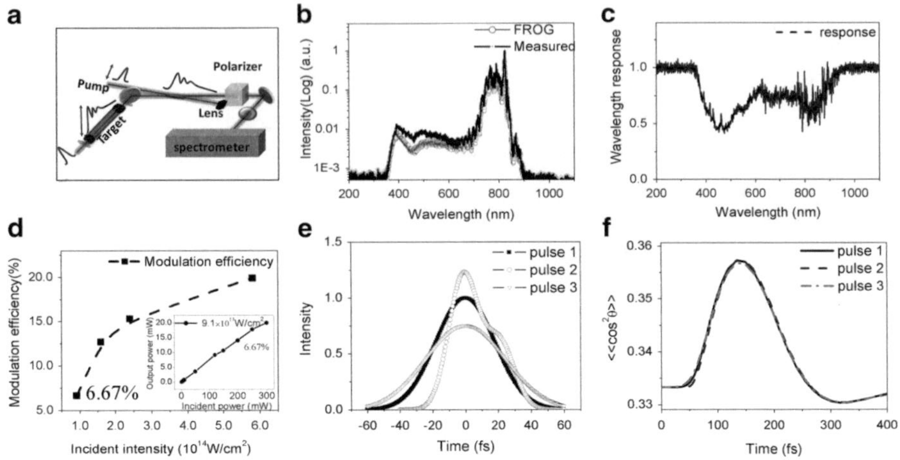

Fig. 3.16 (**a**) Schematic of the SC pulse measurement. (**b**) The directly measured spectrum of the supercontinuum pulse and the time marginal of the M-XFROG trace. (**c**) Wavelength response of the M-XFROG. (**d**) The molecular-alignment-induced polarization rotation efficiency as a function of the excitation intensity. Inset: example power variation of the polarization modulated pulse as a function of incident pulse power. (**e**) Three different excitation pulses with different pulse durations and temporal profiles. (**f**) Simulation of the molecular alignment signal for three different excitation pulses illustrated in (**e**).

the extinction ratio of the polarizer ($\sim$1:10^{-6}), the transmittance of the optical component ($\sim$50%), as well as the modulation efficiency [29]. High sensitivity can be obtained with a high-sensitivity spectrometer, high-transparency optical components, and high-degree aligned molecules.

In a brief summary of this section, a robust FROG technique based on molecular alignment, dubbed as M-XFROG, could be readily realized for full characterization of the electric fields for long broadband supercontinuum pulses, ultrabroadband ultrashort few-cycle pulses, or ultrashort pulses tunable from UV to far infrared. Inherited from its intrinsic mechanism based on linear optical gating with quantum beatings of the impulsively excited molecular rotational wave packets, M-XFROG exhibits various advantages such as linear optical process through polarization rotation with ultrafast molecular gating, "real" gate function as the molecular gating contains no additional phase, applicability to arbitrary wavelength pulses as no phase matching is required, improved sensitivity for weakintensity field characterization, low dispersion in molecular gas as compared to crystals in the standard FROG, low absorption in air for a wide transparent range, convenient performance in air for any wavelength of target pulses, and insensitive to the small fluctuation of the pump pulse. Although the molecular alignment gate is usually a few hundred femtoseconds depending on the molecular species and pumping laser field, such a long gate function works in characterizing ultrashort pulses since a double iterative Fourier-transform algorithm is implemented in retrieving both the molecular gate and target pulse. For a target pulse shorter than the gate width, the molecular alignment gate is effectively sliced by the target pulse. As the time-dependent molecular gate is created essentially on the basis the transient birefringence of the aligned molecules, the gating process takes place by rotating the polarization of the target pulse with a polarization rotation linearly dependent upon the transient molecular alignment. The target pulse transmission through the alignment-induced transient wave plate is thus spectrally resolved versus the time delay with respect to the time-dependent profile of the transient molecular alignment. As we demonstrated experimentally, the M-XFROG could be used to measure few-cycle pulses.

3.4 Ultrafast Optical Imaging by Molecular Alignment

In this section, ultrafast transient molecular alignment is further explored for the ultrafast optical image processing, involving optical image storage in the created molecular wakes and subsequently periodic readout and display. As compared with the well-developed LCD technology [33, 34], this all-optical imaging technology is implemented with quantum wakes of gaseous molecules in air, presenting multiple functions such as the image storage, time-division phase encoding, periodic readout and all-optical display.

From the fundamental point of view, optical image processing with molecular alignment makes use of the alignment-induced spatial cross focusing or defocusing effects, as illustrated in Fig. 3.17a. The aligned molecules change the local refractive index and their spatial distribution enforces an inhomogeneous change of the refractive index, giving rising to spatial focusing (defocusing) for the probe pulse delayed respectively at parallel (perpendicular) alignment revivals. Interestingly, the alignment-induced spatial (de)focusing evolves by closely following the molecular revivals. Using the experimental setup shown in Fig. 3.17b, one can simply monitor the cross-focusing or defocusing of a collinearly propagated probe pulse at different time delays to get a perfect measurement of the alignment matrix. This measurement can readily distinguish the molecular alignment direction (parallel or perpendicular) by measuring the field-free cross-focused or cross-defocused probe pulse intensity around the central part (shown in Fig. 3.17a). Using this technique, we actually get a signal proportional to $(\langle\langle\cos^2\,\theta\rangle\rangle - 1/3)$ rather than $(\langle\langle\cos^2\,\theta\rangle\rangle - 1/3)^2$ obtained in the weak-field polarization spectroscopic method [31]. The detailed description of the direct measurement method and the corresponding experimental results can be found in [31]. As the measurement actually captures the spatial profiles at different delays, i.e., the dynamic evolution of the captured spatial profiles of the probe pulses after the pre-excited molecular wave packets, this technique can be well extended to ultrafast optical image processing. For optical image processing, the gated spatial focusing or defocusing of the probe beam could go a little bit further by writing in the pre-excited molecular rotational wave packets some spatial patterns (revivable

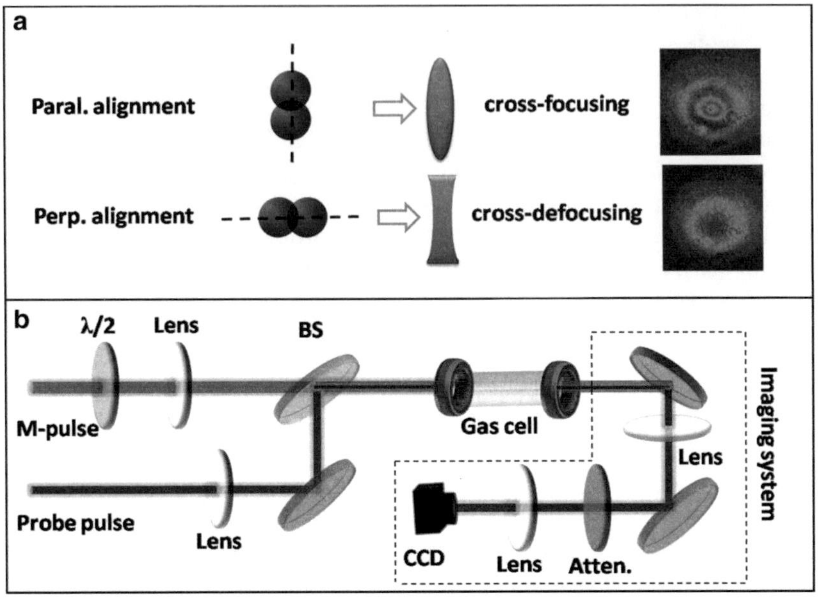

Fig. 3.17 (**a**) Molecular-alignment-induced spatial cross focusing and defocusing. (**b**) The experimental setup for monitoring the cross focusing or defocusing of a collinearly propagated probe pulse at different time delays to get a perfect measurement of the alignment signal

optical image storage) and then reading the storage according to the alignment-induced spatiotemporal modulations (at the molecular revivals).

As schematically depicted in Fig. 3.18, this ultrafast optical imaging technique relies on the weak field polarization technique. A writing pulse (pump pulse) with image information is firstly sent to excite the gaseous molecules and encode the image information in the aligned molecules by creating molecular wakes in space according to the image. After the writing pulse excitation, the pre-excited rotational molecular rotational wave packets of the impulsively excited molecules as storage of image information evolve in free field, presenting periodic revivals of transient molecular alignment. The prealigned molecules act as an ultrafast wave plate by rotating the polarization state of a reading pulse (probe pulse) when the reading pulse time matches the revival of transient molecular alignment. As a result, the image is recovered as the transmission through the polarizer. Moreover, the stored image information can be read out at a desired revival time in respect that the transient molecular alignment revives periodically after the writing pulse excitation. Interestingly, a raised image is obtained at the in-phase revival of molecular wakes based on a cross focusing effect of the parallel aligned molecules while an intagliated image emerges at the antiphase revival of molecular wakes based on a cross defocusing effect of the perpendicularly aligned molecules as shown in the inset of Fig. 3.18.

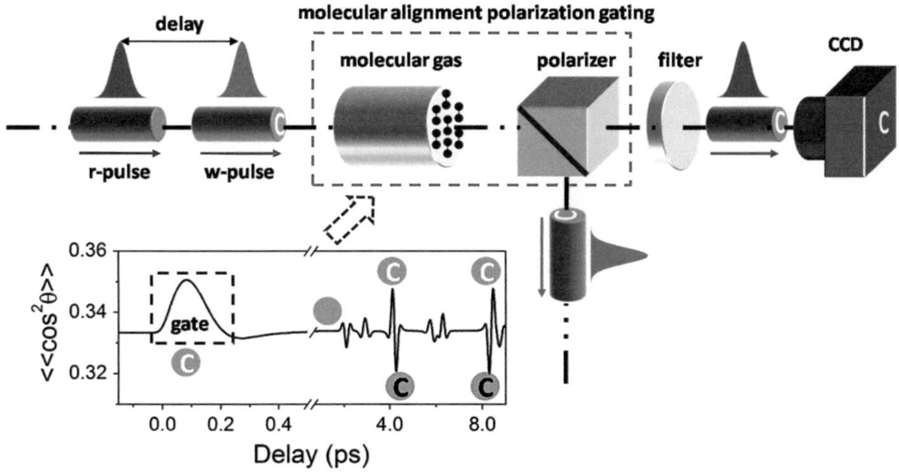

Fig. 3.18 Schematic of ultrafast optical imaging by using molecular-alignment-induced optical gating. A series of predicted output images corresponding to various typical revivals of a calculated molecular alignment are presented in the inset. r-pulse: reading pulse; w-pulse: writing pulse; CCD: charge-coupled device

Experimentally, the writing pulse was at first spatially shaped by passing through a mask with image information and then stored the image information in the created molecular wakes by impulsively aligning the diatomic molecular in air. The polarization of the writing pulse was set to be 45° with respect to the reading pulse. The transmission direction of the polarizer was rotated to be orthogonal to the polarization of the input pulse. Then the succeeding reading pulse was sent to read stored image information from the molecular wakes at a desired time matched to periodic revivals of molecular-induced optical gating. After the polarization modulation by the molecular-induced optical gating, the reading pulse was analyzed with a polarizer, whose transmission direction was first set to be orthogonal to the polarization of the input reading pulse. At the output of the 4f-configuration imaging system, the output images were recorded by CCD. Our experiments were carried out around zero time delay where the molecular alignment signal was strongest due to the combined alignment effects of both nitrogen and oxygen molecules in air.

Figure 3.19 shows the experimental results of the monochromatic optical imaging of the imprinted images. Due to the cross (de)focusing effect [24,35] induced by the transient molecular alignment at the corresponding revival time, both the raised and intagliated images were captured with CCD, which shows a good agreement with the predicted output in Fig. 3.19. Here, images with intensities greater and weaker than the background are presented in white and black colors, respectively.

In order to further develop the ultrafast optical imaging technique based on the molecular alignment, a capacity analogous to holographic imaging [36, 37] was explored to reveal the phase information of a three-dimensional (3D) object. As shown in Fig. 3.20, a transparent glass plate is applied to induce phase infor-

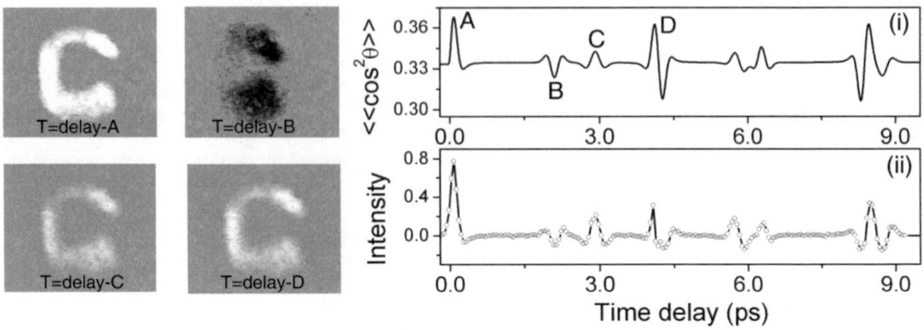

Fig. 3.19 Experimental results of monochromatic imaging by molecular alignment. (**a**) The captured monochromatic images at the time delay when the reading pulse is tuned to match various revivals of the molecular alignment as labeled in (**b**)(i). (**b**) The simulated revivals of the molecular alignment of the diatomic molecules in air (i), and the integrated intensity of the imaged characters versus the time delay of the reading pulse (ii)

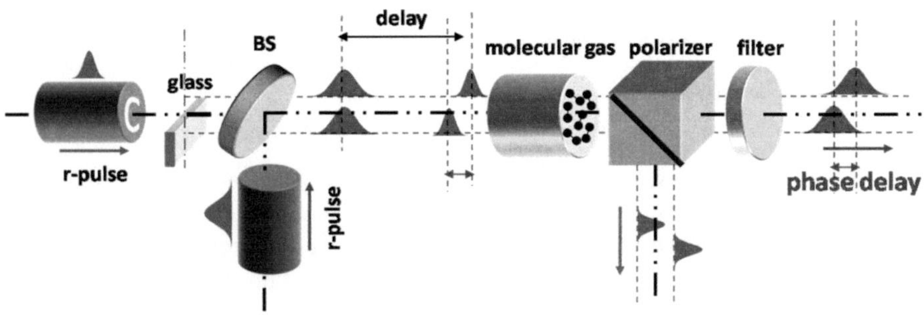

Fig. 3.20 Schematic of the writing and reading schemes for holographic-like imaging

mation of the image so that the image is considered as a 3D one. The writing pulse carries the phase information in the form of a relative time delay between the different spatial parts of beam induced by passing through the glass plate or not. So the phase information is encoded on the aligned molecules by creating spatiotemporal modulations in molecular rotational wakes, realizing a time-division encoding of the phase information of the 3D object.

Figure 3.21 presents an intuitive interpretation of the mechanism of the holographic-like imaging readout. The gray and black calculated molecular alignment signals with a time delay account for the evolutions of the molecular alignment induced by the different parts of the writing pulse with and without passing through the glass plate. So by tuning the reading pulse to match the time delays from the case a to f, the two different parts of the character "C" with and without glass plate are read out by the reading pulse at different time delays. Hence, the phase information can be extracted from the time delay between the cases a and b. In addition, a raised image or an intagliated image is displayed corresponding to its matched revival of transient molecular alignment.

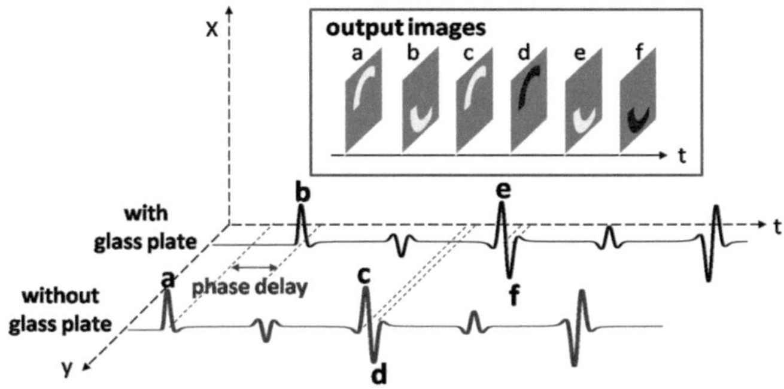

Fig. 3.21 Schematic of holographic-like imaging readout by molecular alignment. The *black and gray lines* are the calculated signal of the molecular alignment with and without glass plate

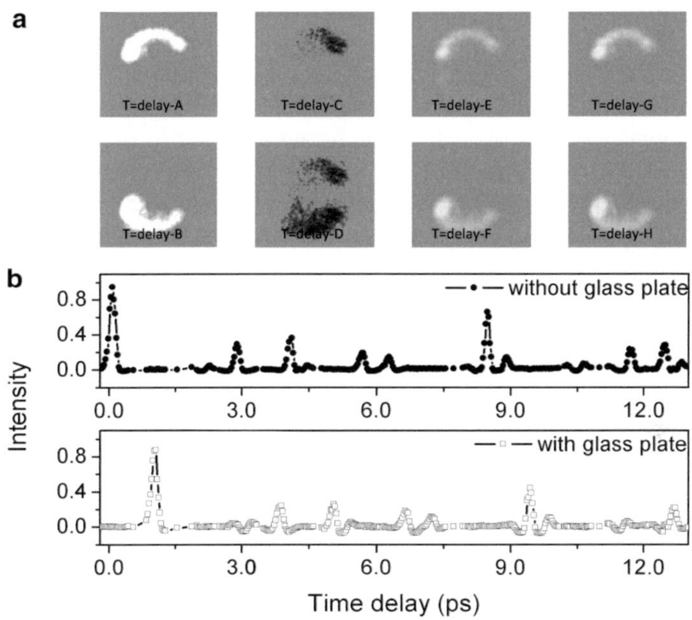

Fig. 3.22 Experimental results of holographic-like imaging by molecular alignment. (**a**) The captured images of a three-dimensional object at various revivals of the molecular alignment [as labeled in (**b**)(i) and (**b**)(ii)]. The phase information is induced by a transparent glass plate after the mask. (**b**) The integrated intensities of the imaged characters versus the time delay of the reading pulse for the object parts without (i) and with (ii) an inserted glass plate

In the experiment, a 500-μm-thick glass plate was inserted as a phase plate to introduce phase information into the image. Half of the writing pulse was modulated by passing through the glass. Figure 3.22 presents the experimental results of holographic-like imaging by molecular alignment. The shift delay induced by inserting the glass plate can be easily obtained as $\sim$ 1 ps from the curve

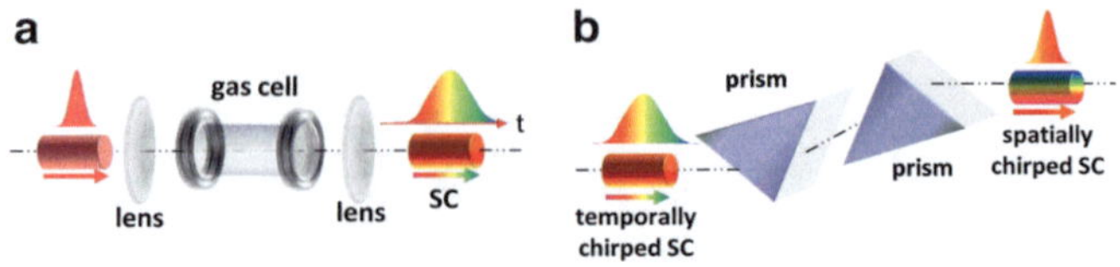

Fig. 3.23 Schematics of the positively chirped SC pulse generation from the gas cell (**a**) and after a chirp compensation by a pair of prism (**b**) in the colorful imaging by molecular alignment.

of molecular revivals, indicating that effective thickness of the glass plate is $\sim 300\,\mu$m. Therefore, by extracting the phase information in a way as the time-phase tomography, this ultrafast optical imaging by molecular alignment exhibits a capacity of 3D imaging.

A colorful display capacity can also be expected of this ultrafast optical imaging by molecular alignment. A supercontinuum pulse with a covering spectral range from near infrared to UV is generated by focusing a weak fundamental-wave pulse into a gas cell filled with atomic Xe at a pressure of 2 atm [38], as shown in Fig. 3.23a, which is applied as the reading pulses. Due to the normal dispersion of the gas and optical components, the red-frequency components of the generated supercontinuum pulse are temporally advancing as compared with the blue ones, i.e., the supercontinuum pulse directly from the gas cell is positively chirped in spectrum. Accordingly, monochromatic images of different colors can be obtained by tuning different temporal slices of the reading supercontinuum pulse to match the transient molecular alignment revivals.

The actual displaying of colorful images required the temporal chirp of the reading supercontinuum pulse to be compensated so that different colors can simultaneously match the molecular alignment. Here, as shown in Fig. 3.23b, a pair of Brewster prisms are used to introduce a negative dispersion and therefore compensate the temporal chirp of the supercontinuum pulse. Above all, either monochromatic displays of different colors or colorful images of spatially resolved colors could be realized by matching the temporally or spatially chirped supercontinuum reading pulse to experience molecular rotational wakes.

Figure 3.24 shows the experimental results of colorful imaging by molecular alignment. The supercontinuum pulse used as a reading pulse covers an ultra-broadband spectral range from 350 to 890 nm, as shown in Fig. 3.24a(ii). Due to normal dispersion, the generated supercontinuum pulse directly after the gas cell is positively chirped. As a result, as presented in Fig. 3.24c(i), (ii), and (iii), monochromatic images in different colors of red, orange, and green could be obtained when the time delay of supercontinuum reading pulse was tuned with its temporal slices of A, B, and C (labeled in Fig. 3.24b) matched the molecular alignment, respectively. By compensating the temporal chirp, the supercontinuum pulse and meanwhile spatially dispersing different colors, as shown in Fig. 3.24c(iv), real colorful images were obtained with colors ranging from orange to green, which accounts for the transmitted spectrum as shown in Fig. 3.24a(ii) (solid

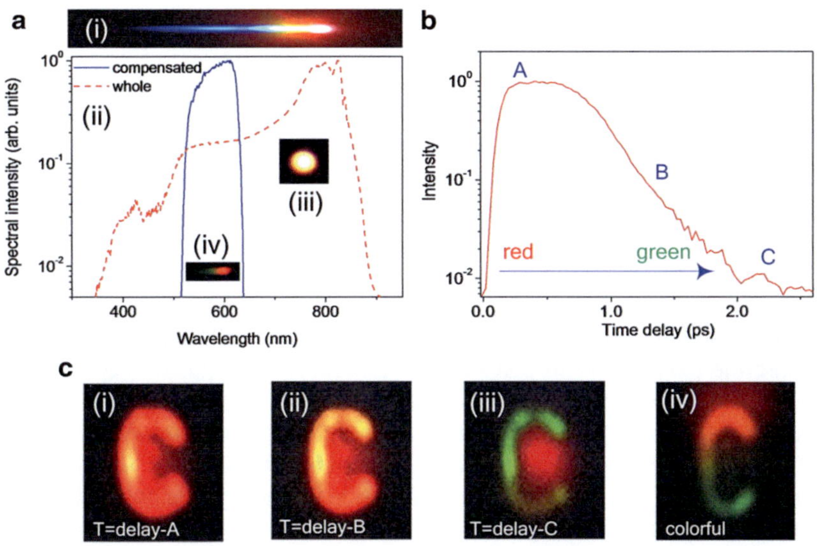

Fig. 3.24 Colorful imaging by molecular alignment. (**a**) The picture of the generated supercontinuum beam profile (iii) and the spatially dispersed one (i). (ii) The spectra of the SC reading pulse with (*solid curve*) and without (*dashed curve*) temporal chirp compensation. (iii) The picture of the spatially dispersed supercontinuum pulse after the Brewster prism pair (see Fig. 3.23b), whose temporal chirp is compensated. (**b**) The temporal profile of the generated supercontinuum pulse from the gas cell with the redder frequency components are advancing with respect to the bluer ones. (**c**) The monochromatic images in different colors when various time slices of the positively chirped supercontinuum pulse [as labeled in (**b**)] is tuned to match the molecular alignment (i, ii, iii). (iv) The real colorful images when a chirp compensated supercontinuum pulse (see Fig. 3.23b) is used as the reading

curve). Furthermore, it is also possible to selectively display the colorful images by manipulating the spatially dispersed supercontinuum pulse at the end of the prism pair, which should be similar to flip the micromirrors for the laser television [39]. With the supercontinuum laser pulse as the reading, the colorful imaging is the real spectral color rather than the conventional red–green–blue mode in the LCD technology.

To sum up for ultrafast optical imaging we demonstrate that molecularalignment-based ultrafast optical imaging and spatiotemporal gating could support robust optical image storage in impulsively excited molecular rotational wave packets. Since optical images encoded in the molecular rotational wave packets could be read out by the use of reading light pulses delayed around the periodic revivals of the molecular alignment, the revivable optical storage equivalently functions as a regenerative optical buffer Note that the optical image storage in the molecular wave-packets could be decoded with ultrashort pulses of different center wavelengths in the spectral range from near infrared to ultraviolet, the readout could display either monochromatically by selecting the reading pulse wavelength or chromatically in real spectral colors by using ultrabroadband supercontinuum pulses rather than the red-green-blue mode in the conventional LCD display. Furthermore, periodic readout of the molecular wakes could be controlled in time and spectral domains to facilitate ultrafast and spectrum-resolved decoding of the optical image storage in the molecular wave packets. Interestingly, by using a spectrally chirped supercontinuum pulse as the reading pulse, the readout exhibits time-coded colors at different delays, and the readout image becomes spatially resolved for different colors if the reading pulse is spatially chirped. Accordingly, we further demonstrate that broadband supercontinuum reading pulses of proper spatial or temporal chirps could support time-gated displays in various colors. As the optical image is stored in the spatial distribution of the impulsively excited rotational wave packets of air molecules, the stored optical images could be spectrally resolved by properly matching the temporally chirped supercontinuum reading pulse to experience the molecular wakes, and a spatially chirped supercontinuum pulse could decode the stored image in spatially resolved colors. In addition to the well-revealed molecular-orbital reconstruction, few-cycle pulse generation, ultrashort pulse compression, and molecular-alignment-based cross-correlation frequency-resolved optical gating, the revivable optical imaging storage in response to the quantum wake of the impulsive rotational excitation is anticipated to stimulate promising applications of optical image processing ultrafast optical buffer, time-encoded and spectral-resolved holographic imaging, as well as regenerative optical storage.

3.5 Conclusions

In summary, we have experimentally demonstrated that femtosecond laserinduced transient molecular alignment with periodic revivals can function as a revivable ultrafast polarization optical gating based on the birefringence of the aligned molecules. Due to the quantum beatings of the rotational wave packets of the impulsively pre-excited molecules, there are periodic revivals of molecular alignment parallel or perpendicular to the polarization of the pump pulse, corresponding to a time-dependent modulation of refractive index. A properly timedelayed probe pulse can experience the ultrafast birefringence effect when propagating through the aligned molecules. Furthermore, the alignment-induced spatiotemporal phase modulation has been investigated including the field-free cross (de)focusing induced by the spatial distribution of molecular alignment and nonlinear phase modulation arising from the time-dependent alignment revivals

A molecular-alignment-based cross-correlation frequencyresolved optical gating (M-XFROG) technique was demonstrated by using the transient molecular alignment as the ultrafast optical gate. A series of measurements of ultrashort pulses with wavelength ranging from UV to infrared were performed by the M-XFROG technique. We have demonstrated that unlike the conventional nonlinear frequency mixing process based FROG geometries, no phase-matching is required by the M-XFROG technique. Hence, the M-XFROG is a robust technique which can be used to characterize ultrashort laser pulses with lowintensity and broadband spectral range

As another application of ultrashort polarization optical gating, we have experimentally demonstrated an all-optical ultrafast imaging technology by using transient molecular alignment. Molecular rotational wakes with periodic revivals were explored for ultrafast optical image processing with multiple functions such as image storage, all-optical display, periodic readout, and time-division phase encoding. It was demonstrated that either raised or intagliated monochromatic images can be obtained at various revivals of time-dependent molecular alignment. Furthermore, both holographic-like imaging which reveals the phase information of a three-dimensional object and a colorful optical imaging by using a supercontinuum laser pulse as the reading pulse are presented.

Acknowledgements This work was supported in part by National Natural Science Fund, and National Basic Research Program of China (2011CB808105). Contributions from H. Zeng's group members and students at ECNU (Jian Wu, Hao Li, Haifeng Pan, and Liang'en Ding) are highly appreciated.

References

1. H. Stapelfeldt, T. Seideman, Colloquium: Aligning molecules with strong laser pulses. Rev. Mod. Phys. **75**, 543 (2003)
2. J.J. Larsen, K. Hald, N. Bjerre, H. Stapelfeldt, T. Seide-man, Three dimensional alignment of molecules using elliptically polarized laser fields. Phys. Rev. Lett. **85**, 2470 (2000)
3. T. Kanai, S. Minemoto, H. Sakai, Quantum interference during high-order harmonic generation from aligned molecules. Nature (London) **435**, 470 (2005)
4. J. Itatani, J. Levesque, D. Zeidler, H. Niikura, H. Pépin, J.C. Kieffer, P.B. Corkum, D.M. Villeneuve, Tomographic imaging of molecular orbitals. Nature (London) **432**, 867 (2004)
5. R.A. Bartels, T.C. Weinacht, N. Wagner, M. Baertschy, C.H. Greene, M.M. Murnane, H.C. Kapteyn, Phase modulation of ultrashort light pulses using molecular rotational wave packets. Phys. Rev. Lett. **88**, 013903 (2001)
6. H. Cai, J. Wu, A. Couairon, H. Zeng, Spectral modulation of femtosecond laser pulse induced by molecular alignment revivals. Opt. Lett. **34**, 827 (2009)
7. J. Wu, H. Cai, A. Couairon, H. Zeng, Wavelength tuning of a few-cycle laser pulse by molecular alignment in femtosecond filamentation wake. Phys. Rev. A. **79**, 063812 (2009)
8. J. Wu, H. Cai, Y. Peng, H. Zeng, Controllable supercontinuum generation by the quantum wake of molecular alignment. Phys. Rev. A **79**, 041404(R) (2009)
9. F. Calegari, C. Vozzi, S. Gasilov, E. Benedetti, G. Sansone, M. Nisoli, S. De Silvestri, S. Stagira, Rotational raman effects in the wake of optical filamentation. Phys. Rev. Lett. **100**, 123006 (2008)
10. S. Varma, Y.H. Chen, H.M. Milchberg, Trapping and destruction of long-range high-intensity optical filaments by molecular quantum Wakes in Air. Phys. Rev. Lett. **101**, 205001 (2008)
11. R. Trebino, K.W. DeLong, D.N. Fittnghoff, J.N. Sweetser, M.A. Krubugel, B. A. Richman, Measuring ultrashort laser pulses in the time-frequency domain using frequency-resolved optical gating. Rev. Sci. Instrum. **68,** 3277 (1997)
12. B. Friedrich, D. Herschbach, Alignment and trapping of molecules in intense laser fields. Phys. Rev. Lett. **74,** 4623 (1995)

13. I.S. Averbukh, N.F. Perelman, Fractional revivals: Universality in the long-term evolution of quantum wave packets beyond the correspondence principle dynamics. Phys. Lett. A., **139**, 449 (1989)
14. P.W. Dooley, I.V. Litvinyuk, K.F. Lee, D.M. Rayner, M. Spanner, D.M. Villeneuve, P.B. Corkum, Direct imaging of rotational wave-packet dynamics of diatomic molecules. Phys. Rev. A **68**, 023406 (2003)
15. I.V. Litvinyuk, K.F. Lee, P.W. Dooley, D.M. Rayner, D.M. Villeneuve, P.B. Corkum, Alignment-dependent strong field ionization of molecules. Phys. Rev. Lett. **90**, 233003 (2003)
16. V. Renard, M. Renard, S. Guérin, Y.T. Pashayan, B. Lavorel, O. Faucher, H.R. Jauslin, Postpulse molecular alignment measured by a weak field polarization technique. Phys. Rev. Lett.**90**, 153601 (2003)
17. V. Renard, M. Renard, A. Rouzée, S. Guérin, H.R. Jauslin, B. Lavorel, O. Faucher, Nonintrusive monitoring and quantitative analysis of strong laser-field-induced impulsive alignment. Phys. Rev. A. **70**, 033420 (2004)
18. J. Wu, H. Cai, H. Zeng, A. Couairon, Femtosecond filamentation and pulse compression in the wake of molecular alignment. Opt. Lett. **33**, 2593 (2008)
19. R. Trebino, D.J. Kane, Characterization of arbitrary femtosecond pulses using frequency-resolved optical gating. J. Opt. Soc. Am. A **10,** 1101 (1993)
20. D.J. Kane, R. Trebino, Characterization of arbitrary femtosecond pulses using frequency-resolved optical gating. IEEE J. Quantum Electron. **29,** 571 (1993)
21. K.W. DeLong, R. Trebino, J. Hunter, W.E. White, Frequency-resolved optical gating with the use of second-harmonic generation. J. Opt. Soc. Am. B **11,** 2206 (1994)
22. T. Tsang, M.A. Krumbügel, K.W. DeLong, D.N. Fittinghoff, R. Trebino, Frequency-resolved optical-gating measurements of ultrashort pulses using surface third-harmonic generation. Opt. Lett. **21,** 1381 (1996)
23. J.N. Sweetser, D.N. Fittinghoff, R. Trebino, Transient-grating frequency-resolved optical gating. Opt. Lett. **22,** 519 (1997)
24. J. Wu, H. Cai, Y. Tong, H. Zeng, Measurement of field-free molecular alignment by cross-defocusing assisted polarization spectroscopy. Opt. Express. **17**, 16300 (2009)
25. K.W. DeLong, R. Trebino, W.E. White, Simultaneous recovery of two ultrashort laser pulses from a single spectrogram. J. Opt. Soc. Am. B **12,** 2463 (1995)
26. P. Lu, J. Liu, H. Li, H. Pan, J. Wu, H. Zeng, Cross-correlation frequency-resolved optical gating by molecular alignment for ultraviolet femtosecond pulse measurement. Appl. Phys. Lett. **97**, 061101 (2010)
27. H. Li, J. Liu, Y. Feng, C. Chen, H. Pan, J. Wu, H. Zeng, Temporal and phase measurements of ultraviolet femtosecond pulses at 200 nm by molecular alignment based frequency resolved optical gating Appl. Phys. Lett. **99**, 011108 (2011)
28. X. Gu, L. Xu, M. Kimmel, E. Zeek, P. O'Shea, A.P. Shreenath, R. Trebino, R.S. Windeler, Frequency-resolved optical gating and single-shot spectral measurements reveal fine structure in microstructure-fiber continuum. Opt. Lett. **27**, 1174 (2002)
29. J. Liu, Y. Feng, H. Li, P. Lu, H. Pan, J. Wu, H. Zeng, Supercontinuum pulse measurement by molecular alignment based cross-correlation frequency resolved optical gating. Opt. Express. **19**, 40 (2011)
30. C. Marceau, Y. Chen, F. Théberge, M. Châteauneuf, J. Dubois, S.L. Chin, Ultrafast birefringence induced by a femtosecond laser filament in gases. Opt. Lett. **34**, 1417 (2009)
31. Y. Feng, H. Pan, J. Liu, C. Chen, J. Wu, H. Zeng, Direct measurement of field-free molecular alignment by spatial (de)focusing effects. Opt Express **19**, 2852 (2011)
32. J. Wu, H. Cai, P. Lu, X. Bai, L. Ding, H. Zeng, Intense ultrafast light kick by rotational Raman wake in atmosphere. App. Phys. Lett. **95**, 221502 (2009)
33. H. Kawamoto, The history of liquid-crystal displays. Proc. IEEE **90**, 460 (2002).
34. J.A. Castellano, Liquid Gold: *The Story of Liquid Crystal Displays and the Creation of an Industry* (World Scientific Publishing Company, Singapore, 2005)
35. V. Renard, O. Faucher, B. Lavorel, Measurement of laser-induced alignment of molecules by cross defocusing. Opt. Lett. **30**, 70 (2005)

36. L. Dhar, K. Curtis, T. Fäcke, Holographic data storage: Coming of age. Nature Photon. **2**, 403 (2008)
37. J. Rosen, G. Brooker, Non-scanning motionless fluorescence three-dimensional holographic microscopy. Nature Photon. **2**, 190 (2008)
38. P.B. Corkum, C. Rolland, T. Srinivasan-Rao, Supercontinuum generation in gases. Phys. Rev. Lett. **57**, 2268 (1986)
39. R. Won, The story behind the screen Nature Photon. **1**, 280 (2007)

Chapter 4
Experiments in Population Trapping in Atoms and Molecules by an Intense Short Laser Pulse

S.L. Chin, A. Azarm, H.L. Xu, T.J. Wang, M. Sharifi, and A. Talebpour

Abstract This chapter discusses some experimental manifestations of interference stabilization or population trapping in atoms and molecules during the filamentation of strong 800-nm femtosecond laser pulses propagating in air and other gases. Particular emphasis is given to nitrogen molecules whose fluorescence induced by the 800-nm pump is probed respectively by 400 nm, 1,338 nm and THz radiations. Fluorescence enhancement and reduction were observed under different probe conditions and at the revival times of the rotational wave packet of nitrogen. Population trapping in the Rydberg states of the molecule is central to the explanation of these observations. We conclude that population trapping through interference stabilization in the multiphoton regime is a universal phenomenon in atoms and molecules in intense laser fields. This includes the excitation of super-excited states of molecules.

S.L. Chin (✉) · A. Azarm · T.J. Wang · M. Sharifi
Center for Optics, Photonics and Laser (COPL) & Department of Physics, Engineering Physics and Optics, Université Laval, Quebec City, Qc G1V 0A6, Canada
e-mail: slchin@phy.ulaval.ca; ali.azarm.1@ulaval.ca; tie-jun.wang.1@ulaval.ca; s-mehdi.sharifi.1@ulaval.ca

H.L. Xu
Center for Optics, Photonics and Laser (COPL) & Department of Physics, Engineering Physics and Optics, Université Laval, Quebec City, Qc G1V 0A6, Canada

State Key Laboratory on Integrated Optoelectronics, College of Electronic Science and Engineering, Jilin University, Changchun 130012, China
e-mail: huailiang@jlu.edu.cn

A. Talebpour
Center for Optics, Photonics and Laser (COPL) & Department of Physics, Engineering Physics and Optics, Université Laval, Quebec City, Qc G1V 0A6, Canada

Qvella, 9133 Leslie St, Suite 110, Richmond Hill, On, Canada L4B 4N1
e-mail: samadt@novxsystems.com

K. Yamanouchi et al. (eds.), *Progress in Ultrafast Intense Laser Science VIII*, Springer Series in Chemical Physics 103, DOI 10.1007/978-3-642-28726-8_4,
© Springer-Verlag Berlin Heidelberg 2012

4.1 Introduction

This chapter attempts to explore and expand on an exceptional idea proposed by Fedorov and co-workers [1–3] in 1988, namely, the interference stabilization of atoms in a strong laser field. Interference stabilization of Rydberg states of an atom results in the trapping of population in some long-lived Rydberg states. Using the idea of dynamic resonance, we observed population trapping in atoms and molecules in 1996 [4, 5]. We try to apply this idea to some molecules in some recent experimental observation of population trapping during filamentation of an intense femtosecond laser pulse in gases, air in particular. One can imagine that a probe laser pulse of different wavelengths would be able to ionize some of these long-lived trapped Rydberg states through one photon absorption. So far, 400 nm, 1,338 nm and even THz (terahertz) pulses were used successfully as probes to ionize the trapped Rydberg states of nitrogen molecules and other gases. Based on our current understanding of population trapping and the super-excited states of molecules pumped by intense femtosecond laser pulses [6–11], we propose that using intense femtosecond laser pulses, the excitation of super-excited states of molecules is mainly due to population trapping. This statement and hence Sect. 4.10 represent our original idea which was not published elsewhere. Consequently, a universal interaction process emerges. When a molecule, no matter how large or small it is, interacts with a short intense laser pulse, there is a probability of trapping some population in some of its Rydberg states and/or super-excited states through interference stabilization.

4.2 Qualitative Physics of Population Trapping

Interference stabilization or population trapping in the context of this work pertains to the interaction of an atom or a molecule with a strong laser field in the multiphoton regime. Interference stabilization was first proposed by Fedorov and Movsesian in 1988 [1]. Essentially, when an atom prepared in a highly excited state (e.g. a Rydberg state) interacts with a short intense laser pulse, there is a probability that the atom resists being ionized and is trapped in a neighbouring Rydberg state in the strong laser field. We refer to Fig. 4.1 for a brief explanation according to Fedorov [3].

An electron of an atom is prepared in a Rydberg level "E_n" ($n \gg 1$). The atom sits in a strong laser field. Ionization takes place by absorbing one photon $h\nu$ into the continuum. However, in the strong broadband radiation (short laser pulse duration), there is a probability that the continuum be coupled to a neighbouring Rydberg state, say, "E_{n-1}" inducing a downward transition through a Raman-type process resulting in a coherent repopulation of Rydberg levels. This up and down transition is called a Λ-type transition. Level "E_{n-1}" could in principle be coupled again to the continuum by absorbing a photon $h\nu$ of the laser field. This would constitute

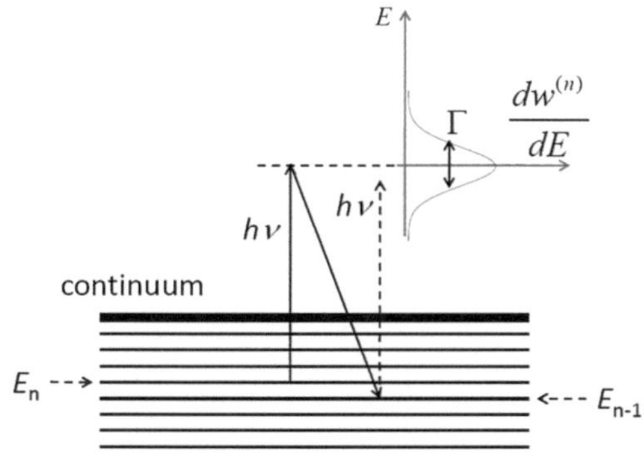

Fig. 4.1 A schematic diagram showing the basic idea of interference stabilization or population trapping à-la-Fedorov [1, 2]

interference of the transitions into the continuum from the two different Rydberg states related to the same electron wave function if the following condition is fulfilled, that is, if the width, Γ, of the ionization probability density curve $\frac{\mathrm{d}w^{(n)}}{\mathrm{d}E}$ (Fig. 4.1) is large enough so that

$$\Gamma \geq E_n - E_{n-1} \tag{4.1}$$

Here, E is the photoelectron energy and $w^{(n)}$, the ionization probability from level "E_n." The width Γ is given by the Fermi golden rule in the ionization rate [3]:

$$\Gamma = \frac{\pi}{2} \left| \vec{d}_{E,n} \cdot \vec{\varepsilon}_0 \right|^2 , \tag{4.2}$$

where $\vec{d}_{E,n}$ is the bound-free matrix element of the dipole moment and $\vec{\varepsilon}_0$ is the field strength of the laser pulse. Under the condition of (4.1), the curves of the ionization probability density, $\frac{\mathrm{d}w^{(n)}}{\mathrm{d}E}$, with different n would overlap and interfere (see Fig. 4.2). It can be seen from (4.2) that Γ is proportional to the square of the laser field strength and hence proportional to the laser intensity. Thus, the higher the intensity is, the larger is the bandwidth Γ. Consequently, the overlap would be larger and it would be more probable that interference of the two transitions occurs. It turns out that the interference is destructive [1–3], resulting in some population being "trapped" in level "E_{n-1}." This means stabilization (or the resistance to ionization).

Before ending this section, it is worth mentioning that the neighbouring levels could be any pair which satisfies the conditions of (4.1) and (4.2). They do not have to be E_n and E_{n-1}. In general, all the Rydberg states would interact with all states of the continuum in the strong laser field [3]. No theory is known yet that gives a clear criteria on the intensity at which trapping would occur or how trapping and ionization compete with each other. However, according to the currently known

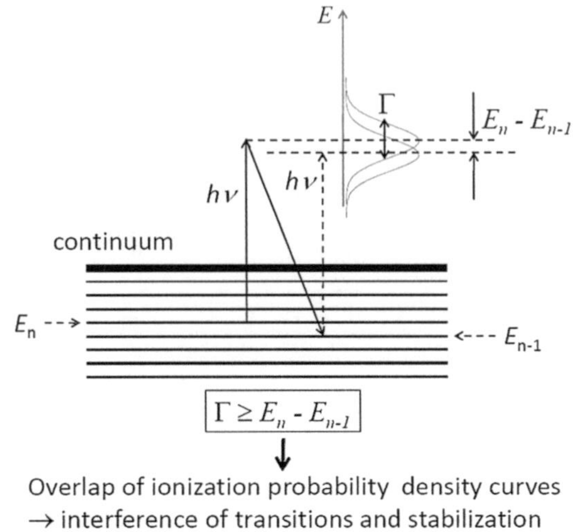

Fig. 4.2 Condition for interference stabilization: overlap of the ionization probability curves

theories [1–3, 15] and experiments [4, 5, 12–14], within a certain intensity range below the saturation intensity, population trapping/interference stabilization would certainly occur. So far, interference stabilization theory has not yet taken into account at higher intensity at which phenomena such as above threshold ionization become important, etc. (Fedorov, private communication, 2011). Thus, through our physical "intuition", at higher intensity, we could say that the coupling of the dynamically resonant Rydberg states with the continuum would become much stronger than the trapping process. This was observed in our previous experiments [4, 5, 14] where the signature of trapping disappears at higher intensity near the saturation intensity (see next section and Figs. 4.4 and 4.5).

4.3 Previous Experimental Observation

The existence of stabilization was experimentally proved in the most unambiguous way by De Boer et al and Hoogenraad et al. in 1993 and 1994 [12, 13] according to the scheme sketched in Fig. 4.1. Ne and Ba atoms were prepared in a Rydberg state and interacted with a strong ps (picoseconds) laser pulse. Resistance of ionization (stabilization) was observed by measuring the decrease in the electron yield when the laser intensity was increased [12, 13]. Later, in our laboratory in 1996, we observed this phenomenon in "an indirect but interesting way" according to Fedorov [3], namely, through multiphoton dynamic resonance and trapping [4,5]. The latter observation and interpretation [4, 5] lead to our current interpretation of new trapping events at high gas pressures. We shall thus give a more detailed description of this latter idea before dealing with the current experimental work.

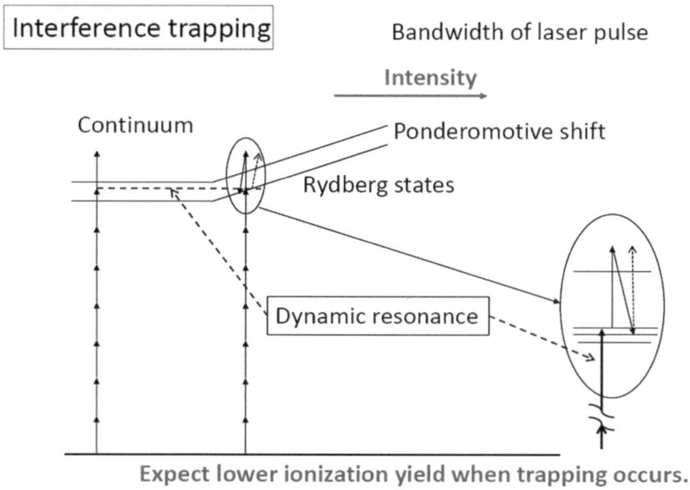

Fig. 4.3 Dynamic resonance induced population trapping

The idea is based upon multiphoton dynamic resonance with Rydberg states of an atom in a strong femtosecond laser pulse (Fig. 4.3). A ground state atom interacts with an intense femtosecond laser pulse in the multiphoton regime (left part, Fig. 4.3). The Rydberg states and the ionization limit will undergo Stark shifts equivalent to the ponderomotive shift of the ionization limit or continuum as the intensity of the laser pulse increases towards its maximum. (Note that the ponderomotive shift is proportional to the intensity.) Some Rydberg states would undoubtedly be shifted into a dynamic multiphoton resonance with the laser field as the intensity increases. This is indicated by the horizontal dashed line linking the end of the next to last photon absorption (left part of the figure) to the shifted Rydberg level in the oval area, which is enlarged in the inset. When this dynamic resonance occurs, a significant population will be excited into the continuum (due to resonant ionization). However, adopting Fedorov's idea of interference stabilization, a downward Λ-type transition in the strong field would favour the trapping of some population into neighbouring Rydberg states. This would invariably occur because the bandwidth of the femtosecond laser pulse is very large so that the ionization probability density curves (see Figs. 4.1 and 4.2) from the two adjacent Rydberg levels (Fig. 4.3, inset) would overlap, resulting in a destructive interference. When this occurs, the ionization probability at this particular intensity would be less than what should have been without trapping. Because there are many Rydberg states below the ionization continuum, dynamic resonance of the laser with these states would occur over a range of intensity. Indeed, we observe this reduction of ionization probability over a range of intensity not only in the case of atoms [4, 5] such as Xe and Kr (Fig. 4.4) but also in simple molecules [4, 5, 14] such as CO and N_2 (Fig. 4.5). It is evident that each ionization yield curve shows some "kinks" indicating trapping at the lower intensity ranges, that is, in the multiphoton regime.

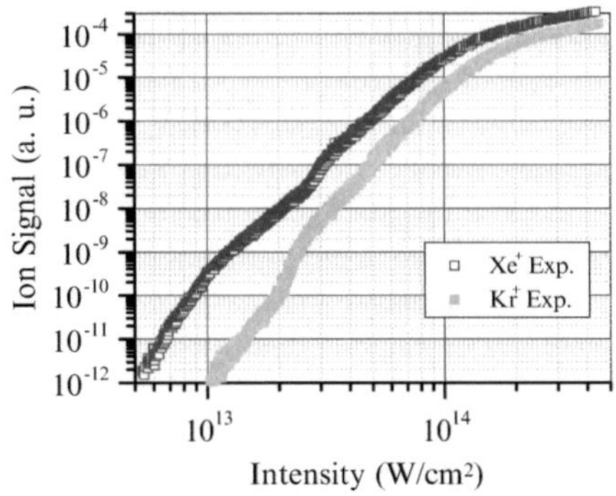

Fig. 4.4 Ion yield versus intensity of laser field for Xe^+ (*open square*) and Kr^+ (*solid square*)

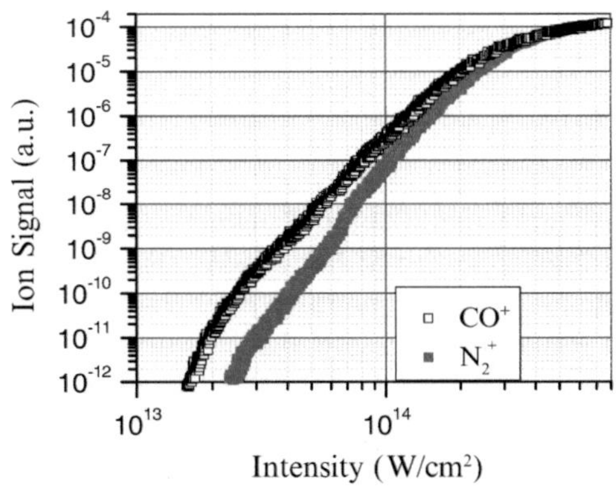

Fig. 4.5 Ion yield versus intensity of laser field for molecular ions CO^+ (*open square*) and N_2^+ (*solid square*)

In all these results, the signature of trapping (slope reduction in the ion yield vs intensity plots) disappears at higher intensities. For example, in the case of Xe (Fig. 4.4), above about 5×10^{13} W/cm^2, the curve becomes smooth, indicating that trapping is reduced significantly. This could be roughly explained as follows. At higher intensities, dynamic resonant ionization would become dominant when competing with trapping. Thus, there is less trapping.

4.4 Physical Picture of Molecular Trapping

We propose in this section a physical picture of population trapping in molecules similar to that in an atom shown in Figs. 4.1 and 4.2. Essentially, we could consider that multiphoton transition in a molecule is instantaneous similar to that in an atom. This means that the transition in a molecule is vertical (Born-Oppenheimer

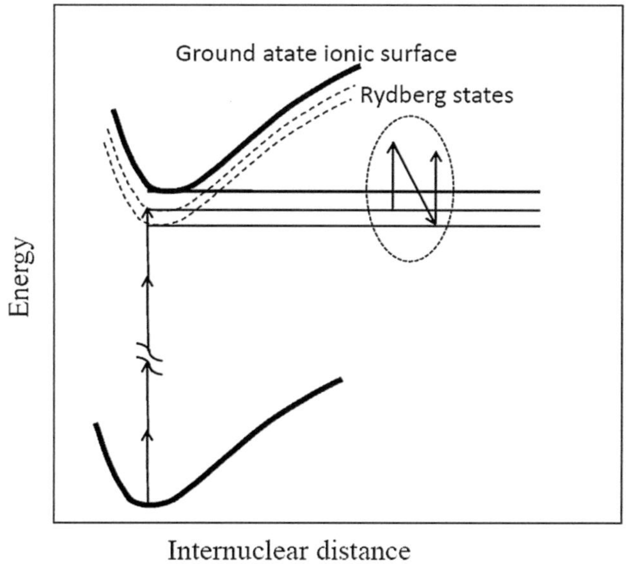

Fig. 4.6 Schematic showing the idea of population trapping taking place during an instantaneous vertical multiphoton transition in a molecule. During the interaction, a dynamic resonance takes place between the ground state and a Rydberg state. From here, trapping occurs in further vertical transition as shown in the *oval area* similar to the case of an atom

approximation) such that the internuclear distance is not yet changed. This is shown in Fig. 4.6. On the left hand side of the figure is shown a multiphoton transition from the ground electronic state towards the ionization limit vertically (at the same internuclear distance). We do not show the ionization channel in the figure for clarity. During the instantaneous vertical multiphoton transition, a dynamic resonance similar to the case of an atom is assumed to have taken place between the ground state and a Rydberg state. From there, trapping in a Rydberg state occurs in further vertical transition as shown in the oval area on the right hand side similar to Figs. 4.1 and 4.2 in the case of an atom.

4.5 More Recent Work in Trapping

Recently, there is a renewed interest in stabilization physics with numerical simulation (see, e.g. [15] and references therein). Another interesting observation is that in the tunnelling regime at higher intensities, trapping was also observed [16–18]. It was called frustrated tunnel ionization [16]. In our recent study of filamentation science in air, we found that population trapping in the multiphoton regime can easily be observed indirectly through dynamic resonances. In what follows, we shall report some recent experimental results showing that trapping of population in the multiphoton regime occurs in an intense femtosecond Ti-sapphire laser pulse. The trapped population is probed by pulses of very different frequencies, namely, terahertz, infrared and visible.

4.6 Probing Population Trapping in Nitrogen Using 400 nm Pulses

Trapping of population of nitrogen molecules by intense near-infrared 800-nm laser pulses was already observed in [12, 13] at very low pressure. We try to probe this in a filament in air at one atmospheric pressure using 400-nm blue laser pulses [19]. Figure 4.7 shows the signal intensity of the fluorescence from (a) the first negative system of N_2^+ ($B^2\Sigma_u^+$–$X^2\Sigma_g^+$ transition) (391-nm band) and (b) the second positive system of N_2 ($C^3\Pi_u$–$B^3\Pi_g$ transition) (337-nm band) as a function of the delay time between the fundamental 800-nm laser pulse and the blue 400-nm probe pulse. Spatial superposition of these two pulses was checked with overlap in the far-field while temporal superposition was checked by observing the diffraction pattern of the 400nm probe beam due to the pump-laser induced plasma. Positive delay time indicates that the probe pulse is behind the pump pulse. The spectra were taken using a spectrometer coupled to an ICCD camera. The data were averaged over 100 laser shots. Because at atmospheric pressure both the lifetimes of neutral and singly charged nitrogen molecules are in the range of hundreds of ps to a few ns [20], the time interval of the ICCD gate was set to 20 ns that is long enough to add up all the fluorescence in the temporal domain. (Note that the ICCD gate delay time was opened 3 ns before the laser pulse arrived at the interaction zone.) It should also be emphasized that when the 400-nm probe pulse is used alone, no fluorescence signal could be detected. It can be clearly seen from Fig. 4.7 that when the 800-nm pump and 400-nm probe pulses are temporally overlapped, that is, the delay time $\Delta t \sim 0$, the fluorescence signals at 391 nm and 337 nm are both increased. It was shown in our previous studies [20] that the 337-nm band is dominantly populated by the dissociative recombination through the following processes: ${N_2}^+ + N_2 \Rightarrow {N_4}^+$; ${N_4}^+ + e \Rightarrow N_2(C^3\Pi_u) + N_2$ at atmospheric pressure. That is, the 337-nm signal is related to the ionization probability, which is the same as the origin of the 391-nm signal; the latter being the result of multiphoton or tunnel ionization of inner-valence electrons of neutral nitrogen molecules [21]. The more ionization there is, the more there will be fluorescence in the 391-nm band and the 337-nm band. Therefore, in the following, we only focus on the fluorescence emission of the $B^2{\Sigma_u}^+$ state of ${N_2}^+$ at 391 nm in order to understand the experimental observation.

The enhancements of the fluorescence emission from the $B^2{\Sigma_u}^+$ state of ${N_2}^+$ at 391 nm can be ascribed mainly to two processes. The first is the two-colour ionization enhancement of ${N_2}^+$ when the pump and probe pulses were temporally overlapped (Fig. 4.8, process b) [14]. The second process is due to the ionization

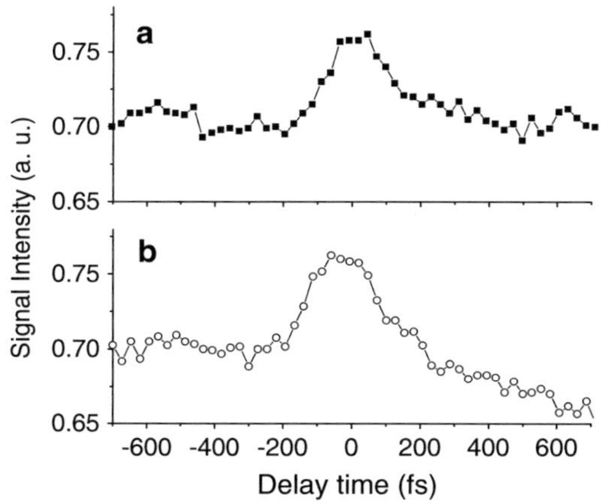

Fig. 4.7 Femtosecond filament-induced fluorescence emission at (**a**) 391 nm and (**b**) 337 nm as a function of time delay between the 800-nm pump and 400-mm probe pulses

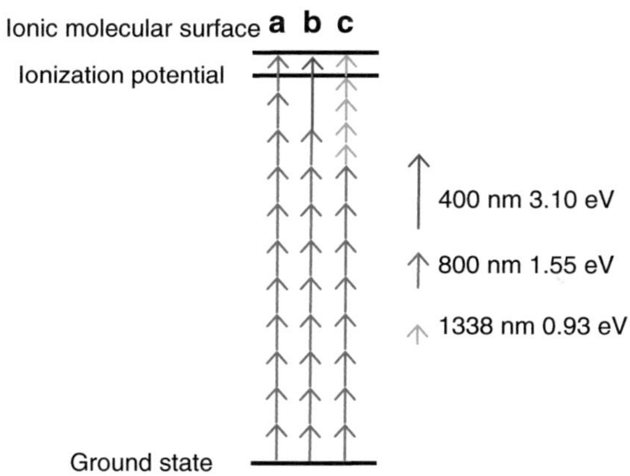

Fig. 4.8 Excitation schemes of nitrogen molecules with (**a**) the 800-nm laser only, and (**b**) the superposed 800 − nm + 400 − nm and (**c**) 800 − nm + 1338 − nm laser pulses

of the aligned nitrogen molecules. Air molecules can be aligned into molecular rotational wave packets by the 800-nm pump pulse [14]. In the case of nitrogen molecules, they take roughly 70–100 fs to make the wave packets aligned parallel to the polarization of the pump pulse. At this time, the pump pulse (50 fs) is gone. When the probe pulse whose polarization is parallel to that of the pump overlaps with the aligned wave packet, enhanced ionization of the trapped Rydberg states would occur, resulting in more ionization and, hence, more fluorescence. The broad width of the enhanced fluorescence peak extending to nearly 200 fs seems to support this delayed fluorescence, which could not be resolved currently because both the pump and probe laser pulses have pulse durations of about 50 fs.

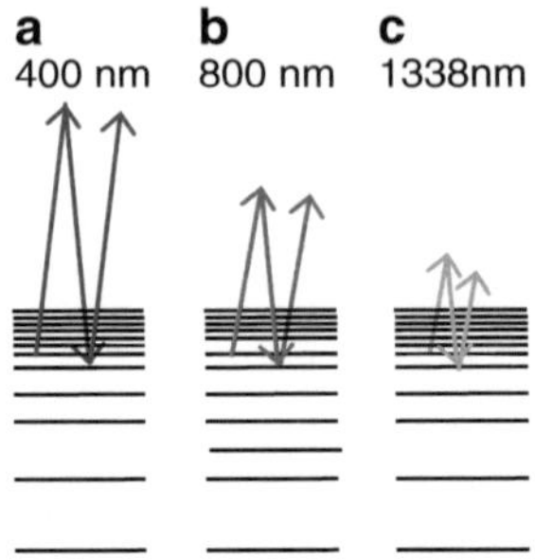

Fig. 4.9 Population trapping induced by (**a**) 400-nm, (**b**) 800-nm and (**c**) 1338-nm laser photons, respectively

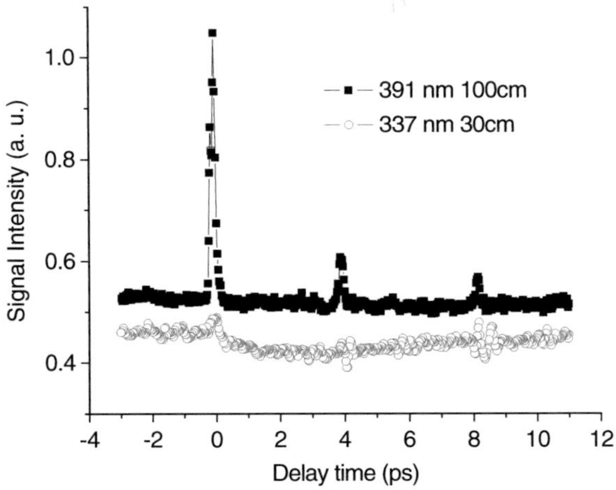

Fig. 4.10 Fluorescence signal of the first negative system of N_2^+ at 391 nm and the second positive system of N_2 at 337 nm versus delay time with corresponding plano-convex lenses whose focal lengths are indicated and positive delay times mean probe pulses are behind pump pulses. The experiment was done in atmospheric pressure air

There would also be contribution of trapping due to the overlapping between the pump and the probe (see, e.g. Figs. 4.3 and 4.6). This would occur when the 400-nm probe induces a dynamic resonance with some of the Rydberg states followed by a Λ-type transition (trapping). However, the latter (Fig. 4.9a) is a higher order process than the ionization enhancement process (Fig. 4.8, process b) and is thus negligible.

Meanwhile, the 800-nm pump could also induce a trapping (Fig. 4.9b). The trapped population could then be ionized by the probe pulse. This enhanced ionization process is of higher order than the direct enhancement by the 400-nm probe (Fig. 4.8, process b), and its contribution could be considered minor.

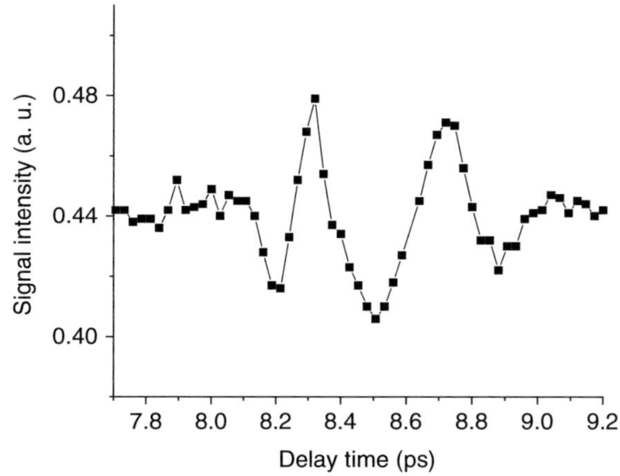

Fig. 4.11 Observed fluorescence signal at 337 nm of neutral nitrogen molecules versus delay time in the vicinity of the full revival period of nitrogen. The focal length of the focusing lens is 30 cm

Moreover, when the pump and probe pulses are well separated by many picoseconds with the probe following the pump, the fluorescence signals at both the 391 nm and 337 nm persist at specific intervals corresponding to the fractional and full revival times of nitrogen's rotational wave packet (Fig. 4.10) [14]. At specific revival time such as the full revival time at 8.4 ps, one can see that the nitrogen fluorescence at 337 nm follows faithfully the rotational change of the wave packet (Fig. 4.11). The reason why the fluorescence follows the rotational wave packet's revival many ps after the pump pulse is long gone is due to trapping. The probe pulse ionizes the trapped states in nitrogen molecules. Between the revival times, there is a background enhancement that does not show any structure. At the revival positions, the probe pulse's linear polarization is alternatively parallel and perpendicular to the aligned molecular wave packets. When they are parallel, the ionization probability is higher [22], resulting in more 337-nm-fluorescence than the background enhancement. When they are perpendicular, the probability of ionization is smaller [22] and the fluorescence is lower than the background enhancement. Another reason for this is that the probe pulse experiences an increase of index of refraction when its polarization is parallel to the aligned molecules, while the index is decreased when the polarization is perpendicular [23]. The index increase (or decrease) would result in more focusing/guiding (or defocusing) of the probe into (from) the rotating molecular zone. This means the probe intensity is higher (or lower) in the filament zone giving rise to more (or less) ionization of the trapped states. In addition, the anisotropy of the emitted fluorescence of molecule with respect to the alignment may influence the signal detection. Considering the nitrogen molecules as radiating dipoles, one expects radiation minimum parallel to the molecular axis and maximum perpendicular to the axis. For randomly oriented molecules,

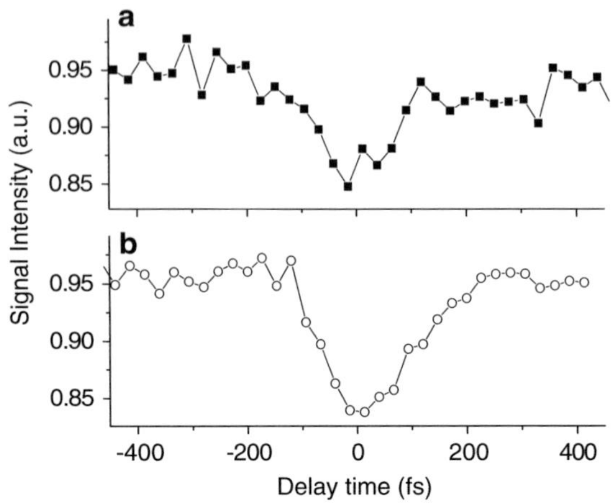

Fig. 4.12 Femtosecond filament-induced fluorescence emission at (**a**) 391 nm and (**b**) 337 nm as a function of time delay between the 800-nm pump and 1330-mm probe pulses

the detected fluorescence signal represents an average. In this experiment, the detection direction is perpendicular to the laser propagation axis and parallel to the pump and probe polarization axis. Hence, the detected fluorescence is minimized when the molecules are aligned and maximized when the alignment direction is perpendicular to the observation axis. These would explain the oscillatory signature of the fluorescence in Fig. 4.10. More detail can be found in [14].

4.7 Probing Trapping in Nitrogen Using 1,338 nm Pulses

Trapping of population of nitrogen molecules by intense near-infrared 800-nm laser pulses in a filament was also probed by using 1338-nm infrared laser pulses [19]. Spatial superposition of these two pulses was checked with overlap in the far-field while temporal superposition was checked by a four-wave mixing (4WM) process between the infrared 1338nm probe beam and the pump: two photons of 800 nm were mixed with one infrared photon ($2\omega_{800} - \omega_{ir}$) leading to a yellowish emission [24]. In this case, as shown in Fig. 4.12, the fluorescence signals at 391 nm and 337 nm are both found to be decreased significantly. The decrease of the fluorescence emission of N_2^+ at 391 nm when applying the infrared 1,338-nm light as the probe is explained by population trapping as described by the processes in Figs. 4.3–4.5. A similar enhancement of ionization process by the 1,338 nm pulse would also occur (Fig. 4.8, process c). However, this being a higher order process than trapping (Fig. 4.9c) could be neglected; that is, its contribution to fluorescence enhancement is negligible while trapping dominates, giving rise to a decrease of the fluorescence shown in Fig. 4.12.

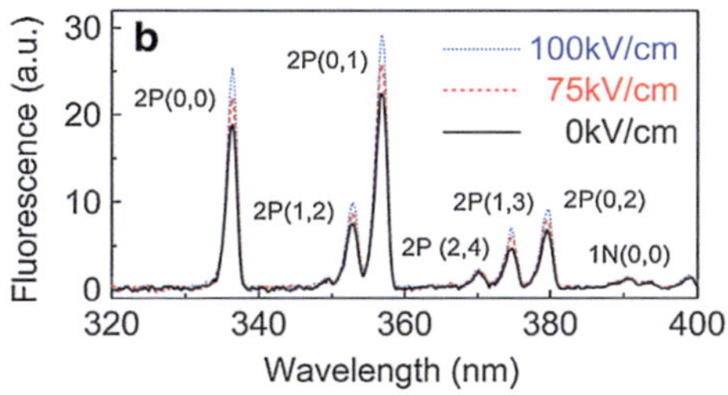

Fig. 4.13 Measured fluorescence spectra versus THz electric field. Major fluorescence lines are labelled. To create plasma, 100 μJ was used. The THz pulse was 1 ps behind the plasma pulse (Adapted with permission from [25] (© 2009 APS), image courtesy of J. Liu, Rensselaer Polytechnic Institute, Troy, New York)

4.8 Probing Trapping in Nitrogen and Other Gases Using THz Pulses

Recently, the group of X-C Zhang has done a series of experiments using THz pulses to enhance the fluorescence from the filaments inside various gases [25–27]. In the case of air, a strong picosecond THz pulse was sent into the filament prepared by a pump pulse at 800 nm (80 fs), and the nitrogen fluorescence (see Sects. 4.6–4.8) from inside the filament was enhanced [25, 26]. The enhancement followed faithfully the change of the THz field giving rise to a new technique for remote stand-off detection of THz pulses in air (Fig. 4.13). This is a manifestation of enhanced ionization by the THz pulse through field ionization of the trapped Rydberg states of nitrogen molecules prepared by the filamenting intense 800-nm pulse.

Further interesting experiment was done by Zhang's group [27] when strong ps THz pulses were used to probe the filament zones pumped by the focused 800 nm/80 fs pulses in inert gases (Xe, Kr, Ar and Ne) at various pressures ranging from one atmosphere to a few tens of Torrs and in some hydrocarbon gases (CH_4, C_2H_6, C_3H_8, C_4H_{10}) at low pressures, a few Torrs. Enhancement of the characteristic fluorescence from all these gases was observed. (We like to point out that, in this series of experiments, there should be no filamentation because the peak power of the pump laser was only about 1 GW or less, lower than the critical power for self-focusing in these gases. But because of this, it gives an advantage to the experiment in the sense that by increasing the pump pulse's energy, the intensity inside the focal zone can be increased because there is no intensity clamping).

In the case of inert gases, the fluorescence is believed to have arisen after ionization and recombination between the electrons and the positive ions in the plasma. The recombination fluorescence would be enhanced if more ionization would occur. The THz pulse was doing just that; that is, ionizing the trapped states of the atoms. What was interesting was that when the laser intensity was increased by increasing the pump energy, the enhancement of fluorescence decreased (Fig. 4.14). This means that there is less trapping when the pump intensity is higher. This was already observed by us [4, 5] and is emphasized at the end of Sect. 4.3.

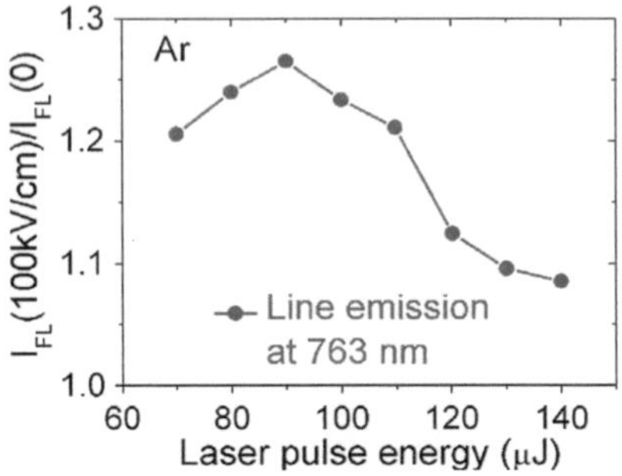

Fig. 4.14 Pulse energy dependence of relative enhancement of Ar 763-nm fluorescence line in the THz field of 100 kV/cm (Adapted with permission from [27] (© 2010 IEEE), image courtesy of J. Liu, Rensselaer Polytechnic Institute, Troy, New York)

In the case of hydrocarbons, the fluorescence from the fragments has been observed by our group in our study of remote sensing of pollutants in filaments [28–30]. In the case of CH_4, we have identified that the fragmentation was due to the excitation of super-excited states of the parent molecules, which "immediately" decayed into the excited fluorescing fragments [7]. Although the precise molecular upper-excited states are not yet known for hydrocarbons, many of the super-excited states of complex molecules are believed to be those Rydberg states converging to excited ionic states of the parent molecules [31]. Thus, the THz pulses might have enhanced the fluorescence of the fragments through ionizing the trapped states, which also decay into the same fragment channels. The above statement is only a speculation. More studies need to be carried out in order to understand the interaction physics.

4.9 Trapping: A Universal Phenomenon for All Atoms and Molecules Including Biomolecules

Summarizing the above experimental observation, evidence of population trapping was shown in some atomic and simple (diatomic) molecular systems. Such trapping takes place principally in highly excited and Rydberg states where the density of states is high as the theory predicted. Probing of the trapped states has been done by measuring the electron yield [12, 13], the ion yield [4, 5, 14] and the high harmonic generation [4, 5] as well as probing the fluorescence yield using probe pulses of very different frequencies (THz, IR and visible) [14, 19, 25–27]. We would deduce that because any atom and molecule, including more complex molecule, possesses highly excited states and Rydberg states, the consequence of its interaction with a short intense laser pulse would result in some trapping (interference stabilization).

The conditions for trapping when an atom or a molecule in the ground state interacts with an ultrafast intense laser pulse are the following. The intensity should be reasonably high so as to be able to pump them into the ionization continuum. Natural or dynamic resonance with some Rydberg states would have occurred. Because the intensity is high, the probability of Λ-type transition would occur, and the width, Γ, of the ionization probability density curve $\frac{\mathrm{d}w^{(n)}}{\mathrm{d}E}$ (Figs. 4.1 and 4.2) is large satisfying (4.1).

Population trapping would, in principle, also occur in bio-molecules. However, in this case, the ionization potentials are rather low. The intensity involved for ionization would thus be low. Using near IR or visible photons, the probability of two or three photon ionization from the ground state and that of Λ-type trapping might be comparable. Ionization might be dominant. Under such conditions, trapping would be difficult to be observed. However, using IR photons, the trapping probability would again be higher because ionization would again involve the absorption of many photons, that is, high intensity.

4.10 Excitation of Super-Excited States Through Trapping by a Strong Laser Field

The idea expressed in this section, as mentioned in the introduction, is original. Super-excited states of a molecule are defined as those neutral states whose energy is higher than the first ionization potential [31]. Using synchrotron radiation, chemists have observed many such states in molecules through single photon excitation [31]. Super-excited molecules will decay through several channels [20], one of which being neutral dissociation resulting in the fluorescence of the dissociated particles [31]. Using intense femtosecond laser pulses, we have observed neutral dissociation of some molecules resulting in their characteristic fluorescence. We experimentally demonstrate that the super-excitation is due to a highly nonlinear process. We proposed in these work that multiphoton resonant absorption into the super-excited states was responsible for this excitation [6–11].

However, let us look closer. Such a resonant absorption in a high-intensity laser field would only give rise to resonant ionization so that there is no more population left in the super-excited state. This is shown in Fig. 4.15 where we assume that the super-excited state is a Rydberg state converging to an excited ionic surface [31]. In the vertical multiphoton transition at the left hand side of the figure, a resonance with a Rydberg state converging to an excited ionic state takes place. Because of the resonance with the Rydberg state, the excited electron would have a high probability to absorb one more photon to the continuum. It will then leave the parent ion with a kinetic energy leaving behind the ion in the excited state. (We do not consider re-scattering because the interaction is in the multiphoton regime, not tunnelling.) No significant neutral Rydberg state would be left behind. However, if the intensity is strong enough satisfying the condition of trapping, a Λ-type transition would occur together with an interference stabilization as shown in the oval area at the right hand side of Fig. 4.15. This trapping (stabilization) process would compete with the resonant ionization process. This is very similar to the trapping during a dynamic resonance with the Rydberg states in atoms and molecules described above. The fact that we have observed super-excitations in many molecular systems experimentally (H_2, NO, O_2, CH_4, etc.) proves that such trapping would have been the excitation process of super-excited states using a short intense laser pulse.

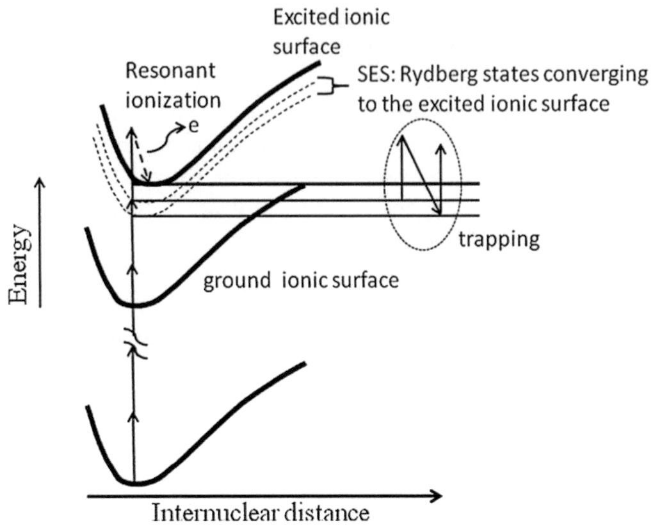

Fig. 4.15 Schematic showing population trapping in the super-excited states (SES). A resonant multiphoton transition into a Rydberg state would have a probability to undergo a Λ-type transition and interference stabilization (trapping). Resonant ionization with the emission of one electron would compete with the trapping/stabilization process

This idea can be further extended to the trapping onto a molecular neutral repulsive surface in the continuum at the initial internuclear distance through vertical transitions. From this position on the repulsive surface, the molecules will immediately undergo dissociation. This seems to be the case in H_2 whose neutral dissociation originates from some neutral excited dissociative surfaces [32].

4.11 Conclusion

We would conclude that population trapping through interference stabilization in the multiphoton regime first proposed by Fedorov et al. seems to be a universal phenomenon in atoms and molecules interacting with an ultrafast intense laser pulse. This would include complex molecules. Population trapping would also be responsible in the excitation of some super-excited states of molecules using intense short laser pulses.

Acknowledgement The discussion with Prof. M. V. Fedorov has enhanced our understanding of the physics of interference stabilization or population trapping. We thank him sincerely for generously sharing his knowledge with us. We acknowledge the technical support of M. Martin. This work was partially supported by NSERC, DRDC Valcartier, Canada Research Chairs, CIPI, CFI, Femtotech and FQRNT, National Natural Science Foundation of China (No. 11074098) and NCET-09-0429.

References

1. M. Fedorov, A.M. Movsesian, J. Phys. B **21**, L155 (1988)
2. N.P. Poluéktov, M.V. Fedorov, J. Exp. Theo. Phys. **90**, 794 (2000)
3. M.V. Fedorov, in *Progress in Ultrafast Intense Laser Science I*, eds. by K. Yamanouchi, P. Agostini, S.L. Chin, G. Farante (Springer, Berlin, Heidelberg, 2006), pp. 1–18. An error in ref. 15 of this cited reference: Should be J. Phys. B 29, 5725 (1996)
4. A. Talebpour, C.Y. Chien, S.L. Chin, J. Phys. B **29**, 5725 (1996)
5. A. Talebpour, Y. Liang, S.L. Chin, J. Phys. B **29**, 3435 (1996)
6. F. Kong, Q. Luo, H.L. Xu, M. Sharifi, D. Song, S.L. Chin, J. Chem. Phys. **125**, 133320 (2006)
7. F. Kong, S.L. Chin, in *Non-Coulomb Explosions of Molecules in Intense Laser Fields*, eds. by K. Yamanouchi, S.L. Chin, P. Agostini, G. Ferrante. *Progress in Ultrafast Intense Laser Science III*, Chap. 6 (Springer, Berlin, Heidelberg, 2008) p. 113
8. A. Azarm, H.L. Xu, Y. Kamali, J. Bernhardt, D. Song, A. Xia, Y. Teranishi, S.H. Lin, F. Kong, S.L. Chin, J. Phys. B **41**, 225601 (2008)
9. D. Song, A. Azarm, Y. Kamali, K. Liu, A. Xia, Y. Teranishi, S. Lin, F. Kong, S.L. Chin, J. Phys. Chem. A **114**, 3087 (2010)
10. K. Liu, D. Song, A. Azarm, S.L. Chin, F. Kong, Chin. J. Chem. Phys. **23**, 252 (2010)
11. K. Liu, D. Song. F. Kong, Laser Phys. **19**, 1640 (2009)
12. M.P. De Boer, J.H. Hoogenraad, R.B. Vrijen, L.D. Noordam, H.G. Muller, Phys. Rev. Lett. **71**, 3263 (1993)
13. J.H. Hoogenraad, R.B. Vrijen, L.D. Noordam, Phys. Rev. A **50**, 4133 (1994)
14. A. Azarm, S. Ramakrishna, A. Talebpour, S. Hosseini, Y. Teranishi, H.L. Xu, Y. Kamali, J. Bernhardt, S.H. Lin, T. Seideman, S.L. Chin, J. Phys. B **43**, 235602 (2010)
15. I.A. Burenkov, A.M. Popov, O.V. Tikhonova, E.A. Volkova, Laser Phys. Lett. **7**, 409 (2010)
16. T. Nubbemeyer, K. Gorling, A. Saenz, U. Eichmann, W. Sandner, Phys. Rev. Lett. **101**, 133001 (2008)
17. N.I. Shvetsov-Shilovski, S.P. Goreslavski, S.V. Popruzhenko, W. Becker, Laser phys. **19**, 1550 (2009)
18. C.I. Blaga, F. Catoire, P. Colosimo, G.G. Paulus, H.G. Muller, P. Agostini, L.F. DiMauro, Nat. Phys. **5**, 335 (2009)
19. H.L. Xu, A. Azarm, S.L. Chin, Appl. Phys. Lett. **98**, 141111 (2011)
20. H.L. Xu, A. Azarm, J. Benhardt, Y. Kamali, S.L. Chin, Chem. Phys. **360**, 171 (2009)
21. A. Becker, A.D. Bandrauk, S.L. Chin, Chem. Phys. Lett. **343**, 345 (2001)
22. I.V. Litvinyuk, K.F. Lee, P.W. Dooley, D.M. Rayner, D.M. Villeuneuve, P.B. Corkum, Phys. Rev. Lett. **90**, 233003 (2003)
23. F. Calegari, C. Vozzi, S. Gasilov, E. Benedetti, G. Sansone, M. Nisoli, S. De Silvestri, S. Stagira, Phys. Rev. Lett. **100**, 123006 (2008)
24. F. Théberge, N. Aközbek, W. Liu, A. Becker, S.L. Chin, Phys. Rev. Lett. **97**, 023904 (2006)
25. J. Liu, X.C. Zhang, Phys. Rev. Lett. **103**, 235002 (2009)
26. J. Liu, J. Dai, S.L. Chin, X.C. Zhang, Nature Photon. **4**, 627 (2010)
27. J. Liu, X.C. Zhang, IEEE J. Sel. Top. Quan. Electron. **17**, 229 (2011)
28. H.L. Xu, J.F. Daigle, Q. Luo, S.L. Chin, Appl. Phys. B **82**, 655 (2006)
29. Q. Luo, H.L. Xu, S.A. Hosseini, J.F. Daigle, F. Théberge, M. Sharifi, S.L. Chin, Appl. Phys. B **82**, 105 (2006)
30. H.L. Xu, Y. Kamali, C. Marceau, P.T. Simard, W. Liu, J. Bernhardt, G. Méjean, P. Mathieu, G. Roy, J.-R Simard, S.L. Chin, Appl. Phys. Lett. **90**, 101106 (2007)
31. Y. Hatano, Phys. Rep. **313**, 109 (1999)
32. A. Azarm, D. Song, S. Hosseini, K. Liu, Y. Teranishi, S.H. Lin, F. Kong, S.L. Chin, J. Phys. B **44**, 085601 (2011)

Chapter 5
Two-XUV-Photon Processes: A Key Instrument in Attosecond Pulse Metrology and Time Domain Applications

P. Tzallas, J. Kruse, E. Skantzakis, L.A.A. Nikolopoulos, G.D. Tsakiris, and D. Charalambidis

Abstract Attosecond pulses today are generated at pulse energies leading to intensities sufficient to induce two-photon transitions in the extreme ultraviolet (XUV) spectral region. Recently, ultra-broadband coherent XUV continua fulfill also the requirements in inducing such processes. Two-XUV-photon ionization is a pivotal tool in attosecond pulse metrology, as well as for XUV-pump-XUV-probe applications targeting the tracking of ultrafast dynamics, providing at the same time spatial selectivity.

Based on these developments, this chapter (a) reviews approaches leading to high intensities of attosecond pulse trains and coherent XUV continua; (b) reviews metrology approaches based on two-XUV-photon ionization, showing their importance through comparative studies with existing XUV-IR cross-correlation approaches; and (c) reports the feasibility of XUV-pump-XUV-probe applications at the 1fs resolution level, in an experiment, where atomic coherence is induced in

P. Tzallas
Foundation for Research and Technology – Hellas, Institute of Electronic Structure and Laser, P.O. Box 1527, GR-71110 Heraklion (Crete), Greece

J. Kruse · E. Skantzakis · D. Charalambidis (✉)
Foundation for Research and Technology – Hellas, Institute of Electronic Structure and Laser, P.O. Box 1527, GR-71110 Heraklion (Crete), Greece

Department of Physics, University of Crete, PO Box 2208, GR71003 Heraklion (Crete), Greece
e-mail: chara@iesl.forth.gr

L.A.A. Nikolopoulos
School of Physical Science, Dublin City University, Glasnevin, Dublin 9, Ireland

G.D. Tsakiris
Max-Planck-Institut für Quantenoptik, D-85748 Garching, Germany

K. Yamanouchi et al. (eds.), *Progress in Ultrafast Intense Laser Science VIII*, Springer Series in Chemical Physics 103, DOI 10.1007/978-3-642-28726-8_5,

© Springer-Verlag Berlin Heidelberg 2012

a rich manifold of doubly excited and inner-shell excited autoionizing states, the evolution of which is tracked through double ionization.

5.1 Introduction

Attosecond (asec) pulse generation [1–3] is maturing following a continuous improvement of the source specifications. In the different campaign of research teams active in *attoscience*, two distinct directions have been followed.

One of those targets records in pulse duration aiming at the improvement of the temporal resolution provided by the attosecond pulses. Enormous progress in this direction has reduced the pulse duration of isolated pulses to 130 asec [4] and more recently to 80 asec [5]. The generation approaches of such pulses are based on few-cycle driving IR laser pulses [6, 7] with stabilized carrier envelope phase (CEP) [8]. The XUV radiation pulses emitted have in the vast majority of the cases low energy of the order of 10–100 pJ. Only recently the Milan team was able to demonstrate 2-nJ pulse energies confined in 150 asec [9]. Asec pulses produced by few-cycle driving pulses, due to their reduced energy, have been used in combination with the driving field, in cross-correlation experimental approaches [1], both in measurements targeting their temporal characterization and in time-resolved applications. The stabilized CEP here is a pivotal parameter as it provides extreme temporal resolution although the laser pulse is in the few fs range. This is because it is not the pulse duration any more but the phase stability that defines the resolution provided by the IR pulse. On cross-correlation measurements are further based on experiments conducted by higher order harmonics and thus XUV pulse trains produced by low-peak-power (pulse energies of the order of one to few mJ), many-cycle, non-CEP-stabilized pulses. Developments in new laser technologies [10, 11] may lead to high pulse energies in few-cycle pulses. Nevertheless, there is so far no experiment demonstrating a nonlinear process induced solely by the XUV radiation of these pulses, which is a natural measure of availability of high peak XUV intensities.

A second direction is mainly targeting high XUV pulse energies leading to intensities allowing for non-linear XUV processes. Two-XUV-photon ionization induced by individual harmonics has been achieved several years ago [12] and exploited in harmonic pulse metrology [13–15] as well as in time domain applications [16]. Two-XUV-photon ionization by a superposition of harmonics has been demonstrated using high-peak-power, many-cycle pulses for the generation of the harmonics [17]. This result has opened up the way to asec pulse metrology via XUV second order autocorrelation approaches. Indeed the first unambiguous quantitative measurement of an attosecond train [18] relied on what is called second order intensity volume autocorrelation (second order IVAC) [19]. This work has been followed by a number of pulse measurements utilizing non-linear XUV interactions such as frequency-resolved two-XUV-photon ionization, XUV-photon above-threshold ionization, two-XUV-photon coulombic explosion [20, 21].

Recently, many-cycle high–peak-power fs pulses have been successfully used in the generation of sub-fs pulse trains emitted by laser surface plasma with pulse energies at the source reaching 40 μJ [22, 23]. The pulse duration was measured in this experiment by the second order IVAC method. Two-XUV-ionization processes have been recently demonstrated to be of increased importance in the metrology of asec pulse trains. A recent comparative work between the second order IVAC and the IR-XUV cross-correlation technique known as "reconstruction of attosecond beating by interference of two-photon transitions" (RABITT) [24] has revealed inconsistencies between the two methods and has shown that the cross-correlation approach may underestimate the duration to be measured, resulting in asec duration even when they are not existing [25, 26]. This result may be subject to the specific conditions of each experiment but the applicability or not of the cross-correlation approach is not a priori known, putting questions on the appropriateness of the approach and its derivatives in asec pulse metrology.

Pulse trains are less appropriate when exploited in the study of ultra-fast dynamics. The generation of energetic isolated pulses is thus a contemporary target in attoscience. Many-cycle high-peakpower pulses are currently used in generating coherent ultra-broadband XUV continua supporting isolated asec pulses. The underlying approach is known as interferometric polarization gating (IPG) [27] that was successfully used in generating sub-100-nJ coherent XUV continua [28]. This new high-intensity source of XUV radiation was very recently exploited in implementing time-resolved measurements of ultrafast evolving induced atomic coherences [29], as well as the first-ever second order IVAC and the firstever 1-fs-scale XUV-pump-XUV-probe [30]. One major drawback when utilizing many-cycle high-peak-power driving IR pulses is the lack of CEP stabilization. This leads to shot-to-shot alternation of the emitted XUV bursts between an isolated pulse and a double peak structure. This has as a consequence the reduction of the fringing contrast in pump-probe experiments and the necessity of the yet not existing single-shot second order IVAC for the measurement of the pulse duration. One way of surmounting this problem is the shot-to-shot tagging of the measured spectra with the value of the CEP. Indeed there are measurable quantities that are sensitive to the CEP value and an approach allowing the on line, shot-to-shot measurement of the CEP value and thus the tagging of the spectra was lately demonstrated [31].

In this chapter we review recent developments in the generation, characterization, and exploitation of intense XUV radiation pulses possibly confined to sub-fs pulse duration. Approaches in reaching high XUV pulse energies are discussed. We further elaborate on the results of the comparative studies between second order IVAC and RABITT and present the recent of the first 1-fs-scale XUV-pump-XUV-probe experiment, in which the evolution of an electron wavepacket formed during the excitation of a dense manifold of doubly excited and inner-shell excited autoionizing states has been tracked.

Throughout the manuscript the laser that has been used is a 10-Hz repetition rate Ti:Sapph system emitting at 800 nm central wavelength pulses of 38–45 fs duration and pulse energy of 170 mJ.

5.2 Sources of Energetic Attosecond Pulses

Since observable non-linear XUV processes require high XUV photon fluxes harmonic generation conversion efficiency has to be optimized. Towards this goal one has to circumvent a number of yield restricting factors. In the next sub-sections we go over the main intricacies of efficiency maximization and address key issues related to XUV sources emitting energetic isolated pulses of ultra-short duration.

5.2.1 Pulse Energy Restricting Factors and Possible Solutions

Optimization of the harmonic yield depends on the driving intensity, phase matching, reabsorption of the XUV radiation, and degree of depletion of the medium. For harmonics emitted by a gas medium the optimal conditions are $L_{\text{med}} > 3L_{\text{abs}}$ and $L_{\text{coh}} > 5L_{\text{abs}}$ [32], where L_{med} is the medium length, $L_{\text{coh}} = \pi/(k_q - qk_0)$ and L_{abs} is the length where the qth harmonic is absorbed by a factor e^{-1}. When the coherence length is longer than both the absorption and medium lengths, the photon yield for the qth harmonic is proportional to $(1-\eta)I_0^n S_{\text{spot}}(PL_{\text{med}})^2$ (where $n \approx 5$ for the harmonics in the plateau spectral region) [33, 34], where P is the gas pressure, η is the ionization probability, and I_0 and S_{spot} are the laser intensity and spot area at the interaction region. However, depletion of the medium through ionization sets a maximum allowed value for the intensity I and the reabsorption for the product PL_{med}. Exceeding the ionization saturation intensity leads to loosing the generating medium and inducing phase mismatch by the free electron cloud. Exceeding the optimum value for PL_{med} leads to yield reduction through absorption. Upon available high laser energies, depletion of the medium can be avoided using loose focusing conditions. Large focal lengths may result to large generation lengths that should be compensated by reducing the atomic density, but at the same time, they result to larger S_{spot} values. Since harmonic generation is a coherent process its yield goes with the square of the number of the emitters and thus is $\propto (S_{\text{spot}})^2$. By using focal lengths of several meters, harmonic pulse energies have reached the 1 μJ level [35]. In all experiments presented in this work, the XUV radiation is produced through harmonic generation in gas targets using "loose focusing" conditions. The yield may be further increased by using quasi-phase matching [36] through multiple gas jets, while pulse shortening increases slightly the ionization saturation intensity.

A promising alternative approach is laser surface plasma harmonics. The plasma medium can in principle not be depleted at whatever laser intensities it is exposed. Indeed the highest harmonic conversion efficiency achieved so far is $6 \cdot 10^{-5}$ (pulse energy 40 μJ) using the plasma medium [22]. The intensity conditions of this experiment were at the limit between coherent wake (CWE) [37–40] and relativistic oscillating mirror (ROM) [41–46] emission. Pure ROM harmonic generation is predicted [42] to reach conversion efficiencies close to 10%. Two-XUV-photon processes induced by surface plasma harmonics are today feasible [22].

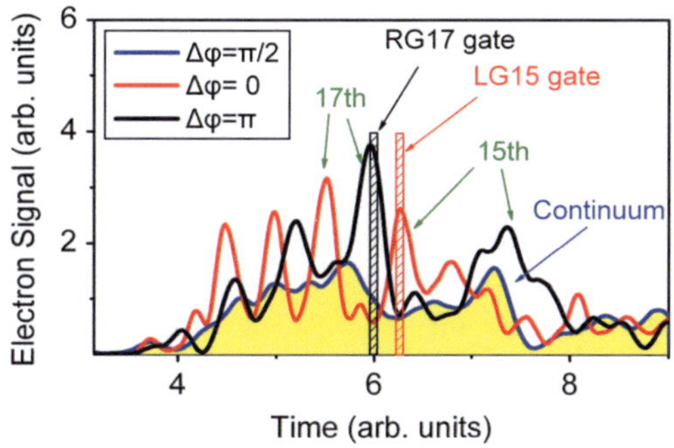

Fig. 5.1 Single-shot XUV spectra recorded in a 500 shots run. The continuum spectrum corresponds to an XUV spectrum generated by a driving laser field with $\Delta\varphi \approx \pi/2$, while the two discrete spectra with a CEP shift $\Delta\varphi \approx 0$ and $\Delta\varphi \approx \pi$. In the present experiment the observed maximum shift of the frequencies ω'_{15} and ω'_{17} is $\omega_x \approx \mp 0.9\,\text{ev}$. $\omega'_{q0} = \omega_{q0} + \varepsilon$ ($q0 = 15, 17$) is the central harmonic frequency which is shifted by ε, as compared to the frequency ωq0 which is produced by a many-cycle laser field. [31]

5.2.2 Intense Broadband Coherent XUV Continua

Optimal conditions for time domain applications of two-XUV-photon processes require energetic isolated XUV pulses. A successful path towards this goal was the development of the IPG technique [27, 28, 47] applied to high-peak-power, many-cycle pulses. The sub-100-nJ coherent broadband continuum emission has been proven sufficient in inducing two-XUV-photon processes. Alternative approaches toward intense isolated asec pulses are based on the wave-plate scheme [48,49], the two-color optical gating [50,51], or the collective effects in XUV generation process [9,52]. Although no two-XUV-photon processes have been so far reported using this technique, they are expected to be soon within reach, at the large advantage of the CEP-stabilized driving laser source used. While IPG is implemented using not CEP-stabilized laser systems, an approach has been developed for the on line, shot-to-shot measurement of the absolute CEP value of the many-cycle driving laser field, which allows tagging of the spectra with the CEP value [31]. The next section goes over this approach.

5.2.3 On the CEP of Energetic 1fs to Sub-fs Scale Pulses

The method [31] is based on the dependence of the XUV spectrum generated by a polarization-gated field on the CEP ($\Delta\varphi$) of the many-cycle laser pulse. In the experiment, the spectral intensity distribution of the XUV radiation generated using an IPG device was determined by measuring the energy-resolved single-photon ionization photoelectron spectra of argon gas. In agreement with calculated spectra, the CEP variation is reflected in the modification of the spectral features (Fig. 5.1), in particular in the frequency shift of the harmonic peaks. The harmonic peak position oscillates between two extremes. These extreme positions correspond to a CEP of 0 and π and vice-versa. This behaviour has also been observed in experiments using few-cycle CEP-stabilized pulses [8,53,54], in which a linear dependence of the peak position shift on the CEP was measured.

The continuum spectrum in Fig. 5.1 is indicative of generation by a laser field with CEP shift $\Delta\varphi \approx \pi/2$, while the other two spectra show harmonic positions undergoing maximum shift, reminiscent of CEP shift $\Delta\varphi \approx 0$ and $\approx \pi$ and vice-versa. In other words, for $\Delta\varphi \approx 0, \pi$, the harmonic maxima are shifting towards their neighboring minima. Aiming at the mapping of the CEP variation of the laser system and evaluating the presence or absence of isolated attosecond pulses in a long-run experiment, two temporal gates of equal width have been set in the time-of-flight XUV PE spectra in order to record the signal (S), which corresponds to the positions of the maximum negative (left) and maximum positive (right) shift of the 15th and 17th harmonics, respectively. The gates are placed on the long electron flight times, where the frequency shift is clearly observable and close to each other in order to have for both gates similar electron collection efficiency and spectral resolution. The signals S_{LG15} and S_{RG17} recorded by the gates LG15 and RG17 are clearly correlated because of the correlation of the harmonic peak position to the shot-to-shot variation of the CEP value. Thus, by placing the gates LG and RG at the frequency positions $(\omega_{\mathrm{RG}(q+2)}, \omega_{\mathrm{LG}q}) = (\omega'_{q+2} - \omega_x, \omega'_q + \omega_x)$ as is shown in Fig. 5.1, and measuring their relative signal the CEP of the driving field can be shot-to-shot retrieved from a parametric plot in an $(x, y) = (\mathrm{NS}_{\mathrm{RG17}}, \mathrm{NS}_{\mathrm{LG15}})$ plane as is shown with filled circles in Fig. 5.2.

In this plot $\mathrm{NS}_{\mathrm{LG15}} = \mathrm{S}_{\mathrm{LG15}}/\mathrm{S}_{\mathrm{LG15(max)}}$, $\mathrm{NS}_{\mathrm{RG17}} = \mathrm{S}_{\mathrm{RG17}}/\mathrm{S}_{\mathrm{RG17(max)}}$ and $\mathrm{S}_{\mathrm{LG15(max)}}$, $\mathrm{S}_{\mathrm{RG17(max)}}$ are the maximum values of the measured signals with $\mathrm{S}_{\mathrm{LG15(max)}} \approx 0.6\mathrm{S}_{\mathrm{RG17(max)}}$. The difference between the amplitudes $\mathrm{S}_{\mathrm{LG15(max)}}$ and $\mathrm{S}_{\mathrm{RG17(max)}}$ allows deduction of the CEP including its sign in the parametric plot. The values of the CEP between the two extreme cases of $\Delta\varphi \approx 0$ and $\approx \pi$, can be retrieved from the ratio $\mathrm{NS}_{\mathrm{RG17}}/\mathrm{NS}_{\mathrm{LG15}}$ (line in the inset of Fig. 5.2). The accuracy with which the CEP value is determined by the accuracy of measuring the ratio $\mathrm{NS}_{\mathrm{RG17}}/\mathrm{NS}_{\mathrm{LG15}}$, which here is $\pm\pi/10$, $\pm\pi/10$ and $\pm\pi/50$ for $\Delta\varphi \approx 0$, $\Delta\varphi \approx \pi$ and $\Delta\varphi \approx \pi/2$, respectively. The squares in the inset are calculated applying the single-atom quantum mechanical three step model. These calculations result in a rather scattered dependence of the harmonic positions on the CEP. To these data a linear fit has been applied [54] in order to extract the smooth curve shown in the inset [31].

Measured points that correspond to a given CEP are lying on lines (dashed lines) that go through the origin. The slope of the line correlates to the CEP value. An important aspect of the parametric map is that it can be used for an online shot-to-shot single asec pulse selection. This can be done by selecting online only those laser shots, which result to a ratio $\mathrm{NS}_{\mathrm{RG17}}/\ \mathrm{NS}_{\mathrm{LG15}} \cong 1$. The procedure is straight forward. Software compares during or after the run the ratio $\mathrm{NS}_{\mathrm{LGq}}/\mathrm{NS}_{\mathrm{RGq+2}}$ of the harmonic q and $q + 2$ signal to a threshold value that warrants presence of

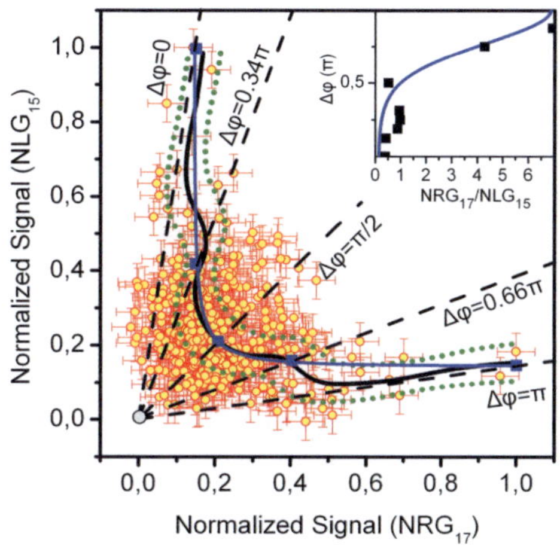

Fig. 5.2 Parametric plot $(x, y) = (NS_{RG17}, NS_{LG15})$ used for the CEP retrieval including its sign. The circles show the correlated NS_{LG15} and NS_{RG17} for consecutive laser shots. The continuous curve is the locus, which depicts the average correlation values of the NS_{LG15} and NS_{RG17} signals, while the dot lines depict the locus of the standard deviation from the average value. The inset shows the dependence of $\Delta\varphi$ on the ratio NS_{RG17}/NS_{LG15}. The squares are calculated using the single-atom quantum mechanical three-step model, while the line (or the line with the dots in parametric plot) is the result of model calculations assuming linear dependence of the harmonic central frequency on the CEP. The dashed lines in the parametric plot show some the indicative values of the ratios NS_{RG17}/NS_{LG15} corresponding to the values of CEP $\Delta\varphi = 0, 0.34\,\pi, \pi/2, 0.66\,\pi$ and π [31]

a continuum spectrum. For ratios exceeding this threshold value, the measured quantities will be disregarded, while for ratios equal or smaller than this value they will be stored.

The method presented can be used as an online shot-to-shot single asec pulse piker since it can select out the pure single *asec* pulses from a series of single- or few-pulse *asec* trains.

5.3 Two-XUV-Photon Processes and Pulse Metrology

Few-photon processes have played a central role in the temporal characterization of ultra-short optical pulses pulse. The simplest approaches rely on two-photon processes. For ultra-short XUV pulses, i.e., for ultra-broadband spectral distributions that can propagate only in rarified media, the only allowed process having a relative small and smooth dependence on the different spectral components is two-photon ionization. The yield of such processes is proportional to $\sigma^{(2)} \cdot (I_{XUV})^2 \cdot \tau_{XUV}$, $\sigma^{(2)}$ being the two photon generalized cross section, I_{XUV} the XUV intensity and τ_{XUV} the pulse duration. Due to the small values of τ_{XUV} (sub-fs) and $\sigma^{(2)}(\sim 10^{-51}\ \mathrm{cm}^4\ \mathrm{s})$, for reasonable atomic densities at the target, I_{XUV} exceeding $10^9\ \mathrm{W/cm}^2$ is required for an observable signal, while for an intensity dependence measurement or a second order autocorrelation intensities higher than $10^{10}\,\mathrm{W/cm^2}$ on target are necessary. This is today feasible for both asec pulse trains and isolated asec pulses. Thus fs pulse metrology approaches based on two-photon processes are extendable to asec pulses at least as far as ionizing intensity is concerned. The next subsections discuss XUV pulse metrology approaches and elaborate on a comparative study between two of those that have revealed the importance of metrology techniques that are based on two-XUV-photon process.

5.3.1 Pulse Metrology Techniques

One of the biggest challenges, that attoscience faces is the development of accurate and reliable temporal and structural analysis tools of the generated asec pulses. The duration τ_{XUV} of these electromagnetic pulses can only be assessed through optical approaches. The characterization of femtosecond laser pulses relies on optical time domain autocorrelation or cross-correlation techniques and frequency domain methods like SPIDER [55]. In attosecond metrology, these methods must be adapted to cope with four critical aspects: the shorter wavelength, the wider spectrum, the shorter pulse duration, and the often much lower pulse energy. Nevertheless, most of the asec pulse metrology approaches are counterparts of the common femtosecond metrology methods. Methods proposed for the characterization of asec pulses include: attosecond streak camera (ASC) [56], which has been successfully used in characterizing isolated asec pulses; the RABITT method, which is the most common approach used for the characterization of low-intensity asec pulse trains; the FROG-CRAB (frequency-resolved optical gating for complete reconstruction of attosecond bursts) [57] method, a common derivative of the RABITT and ASC methods; the XUV-SPIDER [58, 59], which up to now has not been applied to measure attosecond pulses; the in situ [60] method that is seldom used; the second order IVAC [18] and its energy-resolved version [20] applied to intense attosecond pulse trains and very recently to ultrabroadband XUV continua; and the photoelectron analysis with nonresonant two-photon ionization for harmonic electric field reconstruction [20] (PANTHER), a FROG-type approach for XUV pulses has not yet been applied neither for asec pulses trains nor for isolated asec pulses. From the above methods, a comparative study between the RABITT and the in situ method has been implemented in characterizing an asec pulse train [61]. They resulted in different results only at high gas pressures in the harmonic generation region. Availability of intense asec pulse trains has allowed also a comparative study between the RABITT and the second order IVAC method. Since this study is reviewed in the next paragraph we summarize in the following the central intricacies of the two methods.

In the RABITT process photoelectron spectra produced by absorption of photons from harmonics and the driving IR field by an atom, as a function of the delay between the XUV and IR pulses. The spectrum consists of photo-peaks corresponding to the absorption of one photon from each harmonic q, i.e. at positions $E_{main} = qh\nu_L - E_I$, ν_L being the laser frequency and E_I the ionization energy of the atom and additional side bands at positions $E_{side} = (q \pm 1)h\nu_L - E_I$ centered between the main spectral lines of the harmonics. These side bands are the result of different interfering two-photon quantum channels involving absorption of one photon from two subsequent harmonics q and $q+2$ and absorption or emission of one IR photon. Thus the side band signal S_{q+1} oscillates as a function of the delay Δt between the XUV and IR pulses as $S_{q+1} \propto \cos(2\omega_L \Delta t + \varphi_q - \varphi_{q+2} + \Delta\varphi_{atom})$, with φ_L the laser phase, $\varphi_q - \varphi_{q+2}$ the phase difference between the consecutive harmonics and $\Delta\varphi_{atom}$ a known or negligible phase shift induced by the atom. The cross-correlation trace $S_{q+1}(\Delta t)$ of the sideband $q+1$ reveals the phase difference $\varphi_q - \varphi_{q+2} = \varphi_{q+1}$ between two consecutive harmonics. Measurement of the phase differences between all harmonics involved and of their spectral amplitudes the attosecond waveform can be reconstructed as a Fourier synthesis. φ_{q+1} depends on the driving laser intensity. Since the phase φ_{q+1} is not measured by a single-shot method, what is measured is the average phase $<\varphi_q>$ over a number of laser shots of fluctuating intensity and over their full spatiotemporal distribution. This is a central component of the method with important implications in the final duration provided by the method.

The second order IVAC [18, 19] is a more direct assessment of an ultra short light pulse. In the conventional co-linear second order autocorrelation two replicas of an initial pulse, produced by a beam splitter, are non-linearly convoluted by inducing a second order process, the products of which produce a signal $S(\tau) \propto \int_{-\infty}^{\infty} (E_{XUV}(t) + (E_{XUV}(t - \Delta t))^4 dt$, E_{XUV} being the electric field of the XUV radiation. The signal as a function of the delay Δt is the second order AC that reveals to a satisfactory degree the temporal width τ of the initial pulse. In the second order IVAC the beam splitter is replaced by a wave-front dividing split spherical mirror. The delay between the two waves is introduced by translating one of the halves of the split mirror with experimental accuracy of the order of 10 nm. The two halve's of the XUV radiation produce a complex interference field distribution at the focus that is varying with delay. These varying field distributions are two-photon ionizing a gas. The ion signal is modulated with delay resulting to second order autocorrelation trace that has lower contrast (~1:2) than that of the conventional second order intensity AC trace. Since measurement is not single shot, what is measured is the average duration $<\tau>$ over a number of laser shots of fluctuating intensity and over their full spatiotemporal distribution. Thus, a significant difference between the RABITT and the second order IVAC techniques is that the first measures an average phase $<\varphi_q>$, while the second an average duration $<\tau>$. This has consequences with respect to the duration each method delivers. The measured durations by the two methods may differ dramatically. This issue and its origin will be discussed in the next section.

Due to the lack of energetic isolated asec pulses, the second order IVAC has been so far applied only to pulse trains. The newly developed IPG technique, applied to TW many-cycle pulses has produced coherent continua with measured energies and estimated durations sufficient to induce a two-photon process. Indeed, we have recently succeeded in implementing a second order IVAC of such a continuum. Since this experiment was part of the first XUV-pump-XUV-probe study at the 1-fs temporal scale, it will be discussed together in the Sect. 5.5.

5.3.2 Comparative Studies Between the Second Order IVAC and RABITT

Before proceeding with the comparative studies implemented the second order IVAC and the RABITT techniques, it is important to consider what theory predicts on the under different conditions generated pulse durations. Starting with the atomic response, it is well known and pictorially demonstrated by the three-step model [62, 63] that there are two electron trajectories in the continuum, the short and the long ones, that in general contribute to the generation of each "plateau" harmonic. Each trajectory leads to a different time of recombination and thus to a different phase contributing to each harmonic. The saddle-point-solution method [63] results an "inter-harmonic chirp" $\Delta\varphi_q = (\varphi_{q+2} - \varphi_q) - (\varphi_q - \varphi_{q-2})$ that differs in sign for electrons following the short or the long trajectory. The mean phase from the two contributions shows an almost linear dependence with harmonic order.

On the macroscopic scale numerous effects amend the picture providing some control parameters on the efficiency of the generation process as well as on the relative contributions of the two trajectories. To the phase of the generated radiation contribute i) the phase of the driving field, ii) the phase shift due to the atomic (and/or electronic upon strong ionization) dispersion, iii) the geometrical (Gouy) phase shift along the focus, and iv) the atomic dipole phase and thus the variation of the laser intensity in the focal area because of the laser intensity dependence of this dipole phase. The phase matching condition for the qth harmonic reads:

$$\vec{k}_q - q\vec{k}_{\mathrm{L}} = \Delta\vec{k}_{\mathrm{G}} + \Delta\vec{k}_d + \vec{\nabla}\varphi_q$$

The subscripts of the different k vectors denote the laser (L), the Gouy (G) phase shift and the dispersion (d). Since $\varphi_q \propto -t_{\mathrm{e}} I_{\mathrm{L}}$ where t_{e} is the time spent by the electron in the continuum and I_{L} is the laser intensity, $\vec{\nabla}\varphi_q \propto -t_{\mathrm{e}}\vec{\nabla} I_{\mathrm{L}}$. Due to the longer flight time t_{e} of the long trajectory, the phases ϕ^l_q of the harmonic fields emitted from electrons following that trajectory are more sensitive to variations of the laser intensity. Thus the inter-harmonic chirp of the long trajectory fluctuates strongly due to laser intensity fluctuations and/or modulations, eventually preventing attosecond localization [64]. At the same time some of the above contributions can partially compensate for each other favoring "good" phase-matching conditions for the one or the other trajectory. Thus for Gaussian beams according to theory [64, 65], focusing the laser beam almost a Rayleigh range z_{R} before the generation gas provides on axis phase matching for the short trajectory and phase-locking between the harmonics to a degree of more than 50%.

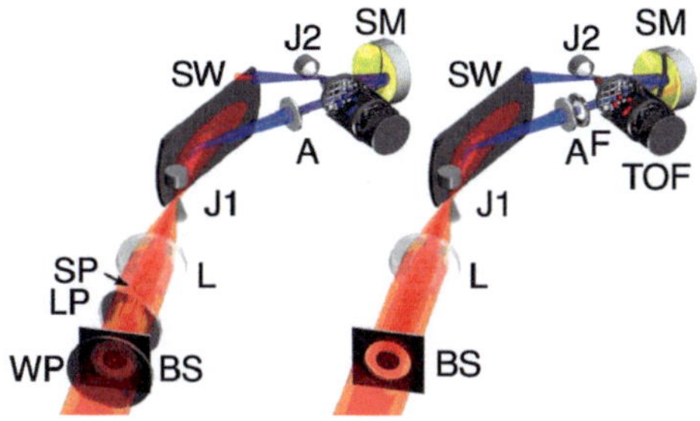

Fig. 5.3 Schematic of the two setups used for the comparative study. The left side illustrates the setup used for the RABITT and the right side for the second order IVAC measurements. A half-WP can be inserted to rotate the polarization of the laser by a small angle. The beam shaper (BS) gives the laser an annular shape, in the case of the RABITT setup it has a hole in the center for the dressing beam. A small glass plate (SP) and a large one (LP) with a hole in the center can be used to add a delay between the central dressing beam and the annular part. The lens (L) focuses the laser at the first gas jet (J1) where the harmonics are produced. The silicon wafer (SW) suppresses the infrared (IR) and reflects the XUV. The aperture (A) blocks the residual IR and lets the central XUV pass together with the dressing beam. A filter (F) can be inserted to select a specific spectral region. The split spherical mirror (SM) focuses the beam into a second gas jet (J2). The time-of-flight spectrometer (TOF) detects either ions or electrons [25]

Focusing the laser beam in the gas jet or almost a Rayleigh range z_R after the jet does not favor phase matching either for the long or the short trajectory. In contrary for a focus almost a Rayleigh range z_R after the jet of axis generation favors phase matching for the long trajectory [64, 65]. For the above reasons attosecond localization was considered to occur only when focusing the beam before the gas jet, i.e. by the short trajectory. Some of the experiments measuring the asec duration took care in fulfilling this condition, while others did not.

The experimental set-ups used for the comparative studies between the second order IVAC and the RABITT techniques are depicted in Fig. 5.3

A second order IVAC of the harmonics generated in the xenon gas jet was taken with the setup depicted in Fig. 5.3 using helium ionization as a nonlinear detector. An indium foil (F) of 150-nm thickness selects the harmonics 9–15 and suppresses the 17th and higher harmonics in order to avoid a single-photon ionization of the helium atoms. The normalized relative amplitudes of the harmonics 9, 11, 13, and 15 were measured to be 100%, 40%, 30%, and 25%, respectively, after reflection from the split spherical gold mirror (SM).

Three autocorrelation traces measured at the three different positions of the focusing lens (L) are shown in Fig. 5.4 One with the focus at position $z_B = -0.86 z_R$ before the xenon gas jet (J1), one with the focus at position $z_I = 0$ inside the jet, and one at position $z_A = +0.56 z_R$ focusing after the gas jet. The length of the Rayleigh range of the focus is $z_R = 88 \pm 5$ mm.

Only if the focus of the driving laser is before the harmonics generating gas jet at z_B, then the second order IVAC trace in Fig. 5.4 shows clear and regular periodic structure with maxima appearing twice per laser period T_L. The solid line in the top

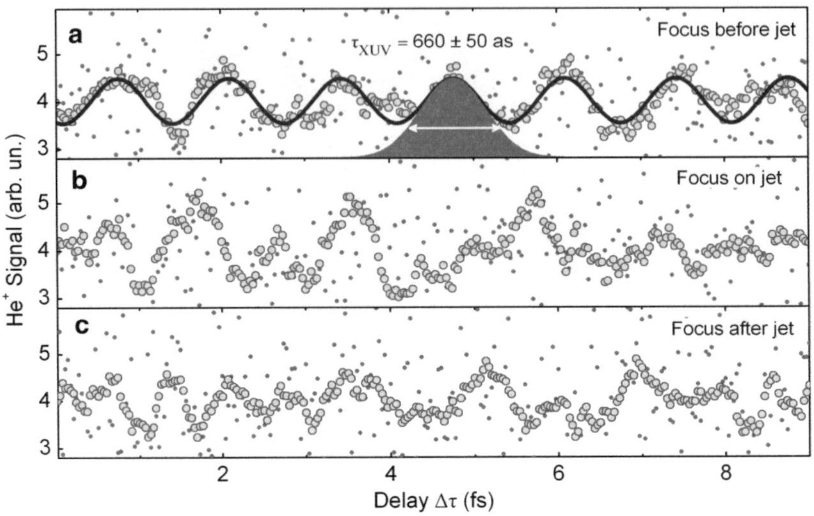

Fig. 5.4 (**a**)–(**c**) 2-IVAC traces recorded for the focus positions z_B, z_I and z_A. The *dots* are the measured data. The *circles* correspond to a 10 point running average. The *line* in (**a**) is a 12-peak sum of Gaussians fit to the raw data. The *filled area* in (**a**) is one of the 12 Gaussian pulses of the fitted function [25]

panel of Fig. 5.4 shows a best fit to the data for a series of Gaussian peaks with a fixed spacing of half a laser period. From the width of the Gaussians the average pulse duration of the pulses inside the train was determined to be $\tau_{XUV} = 660 \pm 50$ asec. The traces recorded with the IR driving laser focused inside or after the gas jet show merely random fluctuations and no regular structure. This is expected from theory behaviour as discussed above.

Most striking results of this work are the RABITT traces obtained at the three different focusing positions before z_{BJ}, inside z_I, and after z_A, the first gas jet (J1), as shown in Fig. 5.5.

For the cross-correlation between the infrared and the harmonic radiation, a setup very similar to the one applied by Paul et al. [24] is used with the beam shaper (BS) shown in Fig. 5.3 having a hole in the center to form the small central dressing beam. The intensity of the IR dressing beam can be adjusted by rotating a $\lambda/2$ half-wave plate (WP) placed before the BS by a small angle. Two plates provide the delay Δt between the central IR dressing beam and the XUV. One large plate (LP) with a hole in the center delays the annular IR driving laser field and one small plate (SP) delays the central dressing beam. Rotation of the central plate (CP) by a small angle changes the delay between the central and the annular part of the IR and thus the delay Δt between the dressing IR and the XUV produced from the annular beam. Figure. 5.6 shows a typical photoelectron-energy spectrum of argon ionized by the superposition of the dressing IR beam and the harmonics taken while scanning the delay Δt to obtain the RABITT traces.

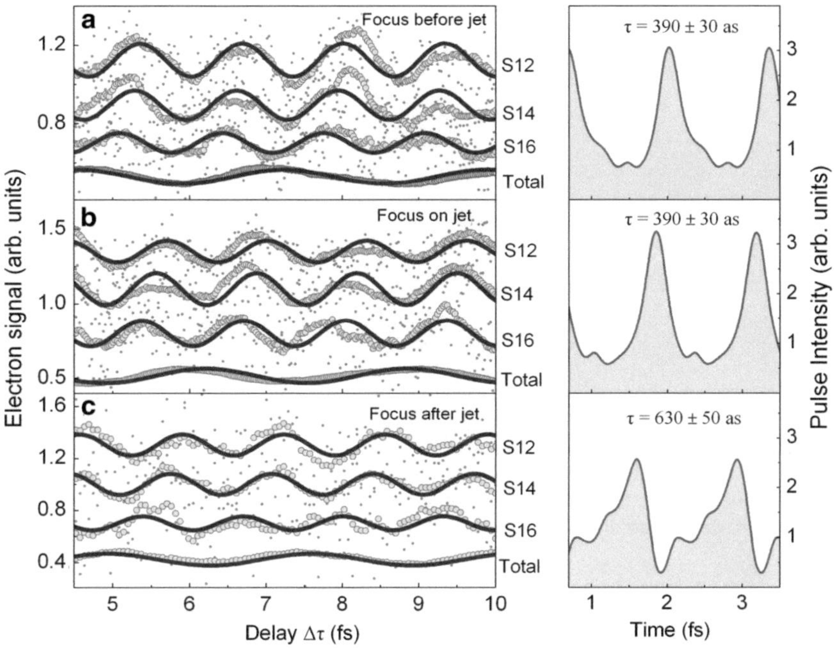

Fig. 5.5 (*Left panel*) RABITT traces measured at the three focus positions normalized to the corresponding total signal. The *dots* are the measured data. The *circles* correspond to a running average over 15 points for the sidebands and 40 points for the total signal. The *purple lines* are sinusoidal fits to the raw data over 13 oscillations on the sideband traces and 6 laser oscillations the total signal. (*Right panel*) Reconstructed pulse trains [25]

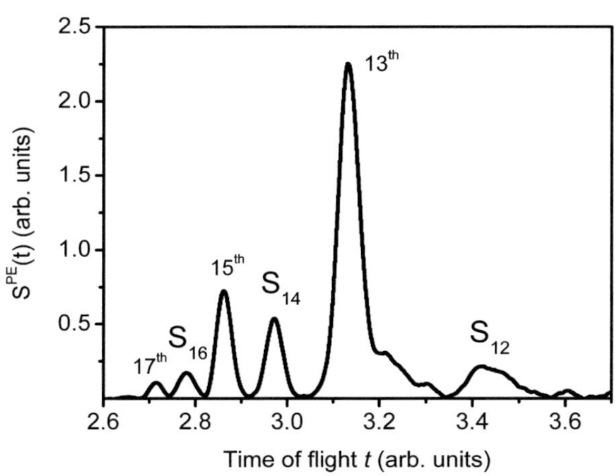

Fig. 5.6 Measured time-of-flight photoelectron spectrum with side bands. The photoelectron energy spectra $S^{PE}(t)$ of argon ionized in the presence of both the XUV harmonics and the fundamental IR show the side-bands S12, S16, and S18, appearing centered between the odd harmonics

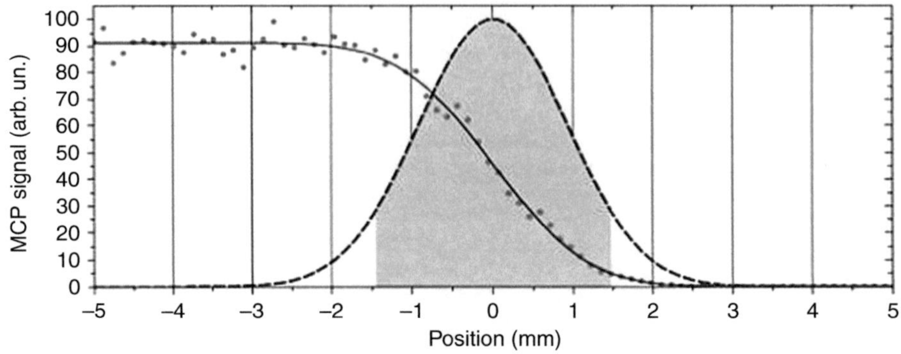

Fig. 5.7 Measurement of the harmonic radiation diameter. The edge of a knife was inserted in order to block a part of the XUV beam. Recording the photoelectron spectra as a function of the knife's position reveals the diameter d_{XUV} of the harmonics. The *gray dots* are the measured photoelectron signals. The *solid line* is an error function fit to the measured data. The *dashed line* is the derivative of the fit. The *gray area* illustrates an aperture of $d_{\mathrm{A}} = 3$ mm. In this measurement the laser focus was before the gas jet

The results of the RABITT measurements shown in Fig. 5.5 imply almost Fourier-transform limited (FTL) attosecond pulses of τ_{XUV} 390 as even when focusing the IR beam inside the harmonics generating gas jet. Adjusting the phases φ_q of the harmonics extracted from the RABITT measurement according to the dispersion of the 150-nm-thick indium filter used in the second order IVAC measurement and using the same relative amplitudes for a direct comparison results in the durations τ_{XUV} of the reconstructed pulses of 350, 600, and 1,020 asec for the focus positions z_{B} before, z_{I} inside, and z_{A} after the xenon gas jet, respectively.

This is in contradiction to both theoretical predictions and the second order IVAC results shown above, where asec confinement was observed only for the focus position before the jet. This behavior originates from the presence of both the long as well as the short electron trajectories, producing the harmonic radiation, although with different weight, in all three focus positions [25,26]. As experimentally verified the long trajectory is not fully eliminated for a focus after the jet, despite the inserted aperture with diameter of $d_{\mathrm{A}} = 3$ mm, which selects only the central harmonic beam. This aperture blocks half of the total radius of the XUV beam produced when the focus of the driving laser is placed before the gas jet. This is shown in the knife edge measurement of Fig. 5.7. Note that placing the focus after the jet the XUV diameter is substantially increased [66, 67], thus the aperture is blocking a much larger fraction of the beam.

A further relevant measurement is the total XUV signal measured as a function of the position of the focus. The results are shown in Fig. 5.8. As can be seen despite the use of the aperture that blocks the largest part of the radiation produced by the long trajectory, the signal for the focus after the jet is reduced at most by only 20%. No use of an aperture results in a much higher signal when the focus is placed

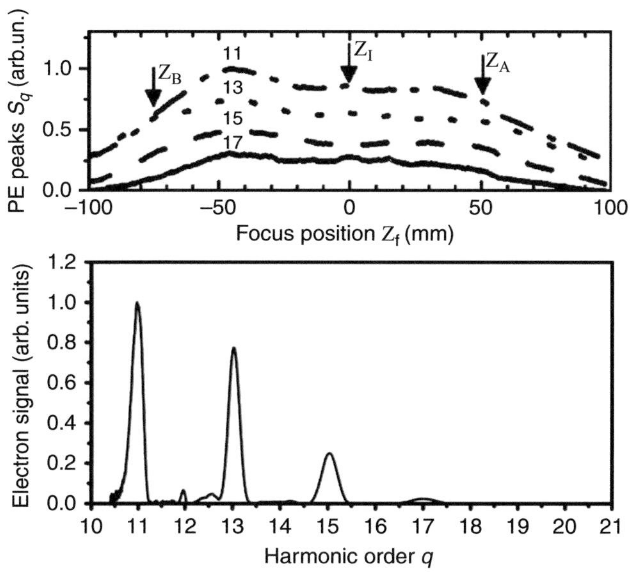

Fig. 5.8 Harmonic generation at different focus positions. The intensities of the individual harmonics depend on the position of the focus relative to the xenon gas jet. The *top panel* shows the photoelectron (PE) signals S_q as a function of the focus position z_f for the harmonics 11, 13, 15, and 17 by the *dash-dotted*, the *dotted*, the *dashed*, and the *solid line*, respectively [25]. The *bottom panel* shows a recorded photoelectron-energy spectrum with rescaled abscissa showing the peaks from the harmonic orders 11, 13, 15, 17, and 19. The spectra were recorded with an aperture of $d_A = 3\,\text{mm}$

after the jet indicative of more efficient harmonic generation by the long trajectory. The higher conversion efficiency of the long trajectory results into reduced but non negligible contribution to the central part of the XUV radiation.

As discussed above, RABITT measures an average harmonic phase $<\varphi_q>$. The averaging involves long and short trajectory contributions, contributions from the spatial intensity distribution, of the temporal distribution and the shot-to-shot laser intensity fluctuations. This averaging results into a smoothening of the spectral phase and amplitude distribution measured, leading to almost zero chirp and smooth spectrum, i.e. to an artificially deduced almost Fourier-transform-limited pulse. Thus, at least for the conditions of the present experiment, the RABITT technique and techniques based on it (e.g. CRAB-FROG of pulse trains) strongly underestimates the measured duration and infers attosecond confinement [24] even when it is not existing [25].

The results of this comparative study introduce an ambiguity with respect to the applicability of some of the attosecond metrology approaches in previous and future work. This question may be answered by extending this type of studies to different experimental conditions, such as different spectral regions and different focusing conditions. Work in this direction is in progress.

5.4 XUV-Pump-XUV-Probe Experiments at the 1-fs Scale

While non resonant two-XUV-photon ionization is the basis for a second order IVAC, its resonant counterpart serves in studying the dynamics of the intermediate state. This is the principle of the pump-probe approach, in which one pulse is exciting an evolving wavepacket, while a second delayed pulse is probing the evolution e.g., by ionizing the system. The cleanest pump-probe scheme is that using two different colors for the two pulses. This is because the use of a single color introduces background to the signal originating from the two-photon absorption from each of the two pulses. Two-color experiments in the XUV spectral region with sub-fs pulses are not yet realized. Nevertheless the first single-color XUV-pump-XUV-probe experiment at the borderline between the femto—and attosecond regimes has been recently achieved, using energetic coherent continua produced by the IPG technique [27]. An atomic coherence was induced in the vicinity of doubly excited and inner-shell excited autoionizing states of xenon, through single-XUV-photon absorption. The evolution of the atomic coherence was then tracked through absorption of a second XUV photon from a delayed pulse that led to double ionization of the atom. Interestingly, the same trace at delays close to zero contains the second order IVAC of the radiation used. This is because due to the shortness of the pulse used, direct double ionization becomes more efficient than the sequential one [68]. This new physics and the achieved measurements will be summarized in the following sections.

5.4.1 The Two-XUV-Photon Double Ionization of Xenon Scheme

For the multi-XUV-photon ionization of xenon has been used XUV continuum radiation produced through the experimental setup shown in Fig. 5.9a.

The laser beam was passing through a double Mach-Zehnder interferometric polarization gating (DMZ-IPG) device. The ellipticity-modulated laser beam after exiting the DMZ-IPG was focused with an $f = 3$ m lens into a pulsed xenon gas jet (GJ1), where the XUV broadband coherent continuum radiation was generated. After the jet, a Si plate was placed at Brewster's angle of 75° of the fundamental, reflecting the harmonics toward the detection area, while substantially reducing the IR field. The XUV radiation after reflection from the Si plate passes through a 3-mm-diameter aperture and a 150-nm-thick Sn filter in order to select the central part of the beam, and the required spectral region with a central wavelength of $\approx$60 nm (Fig. 5.9b), as well as to eliminate the residual part of the IR beam. Subsequently the XUV pulse was split in two halves and focused into the target xenon gas jet (GJ2) by a split spherical gold mirror of 5-cm focal length. Ion mass spectra were measured with a time-of-flight spectrometer. One spectrum is shown in Fig. 5.9c in which the $Xe^{+}$and Xe^{2+} ion peaks are clearly seen. The autocorrelator delay was

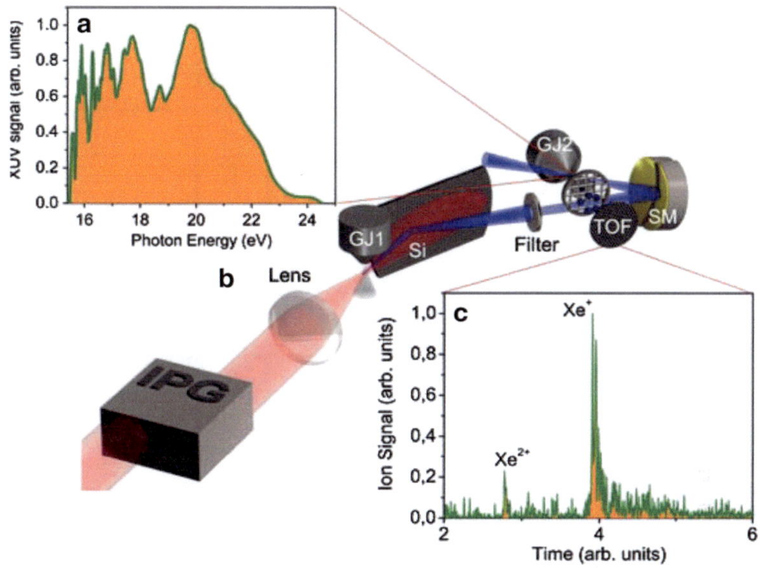

Fig. 5.9 (**a**) Experimental setup. IPG: interferometric polarization gating device; GJ1,2 Xenon gas jets; TOF: time of flight spectrometer; SM: split spherical mirror. (**b**) Emitted continuum radiation (averaged over several pulses). (**c**) Ion mass spectra in which Xe^{+} and $Xe^{++}$are observable [30]

introduced by translating one of the two halves of the split mirror. The minimum displacement step of the z-translation piezo-crystal unit is 1.5 nm. The spectral region from ≈80 nm to ≈50 nm (Fig. 5.9b) was selected by placing the Sn filter in the XUV beam. The double ionization scheme that was studied is depicted in Fig. 5.10. For the given XUV spectral width Xe^{2+} can only be produced through multi-XUV-photon absorption. The most pronounced possible channels are: (a) two-photon direct double ionization (TPDDI). Here single XUV photon absorption is in the vicinity of a manifold of doubly excited $5s^2 5p^4[^3P,^1D,^1S]$ *mlm'l'* and inner-shell excited $5s5p^6(^2S_{1/2})np[^1P_1]$ (Auger) autoionizing states (AIS). Absorption of a second XUV-photon ejects two electrons that share the excess energy, and (b) Single-XUV-photon ionization of Xe, followed by a single-XUV-photon ionization of the two fine structure levels of the Xe^{+} ground state by the high energy (above 21 or 19.7 eV for each state) part of the continuum spectrum distribution, leading to sequential double ionization (SDI). Additional less probable sequential channels may include in some runs single-photon ionization of xenon leading the ion in its lower-lying excited states. While for pulse durations τ_{XUV} of tens of fs the sequential process is the dominant one, because of its higher cross section, the direct process gains in relative strength as the pulse duration decreases. This is because of the different dependence of the two processes on the pulse duration. The direct one is $\propto \tau_{XUV}$ and the sequential $\propto \tau_{XUV}^2$ [68, 69]. At the 1 fs pulse duration level the direct process becomes the strongest double ionization channel, verified experimentally in this work. Using a

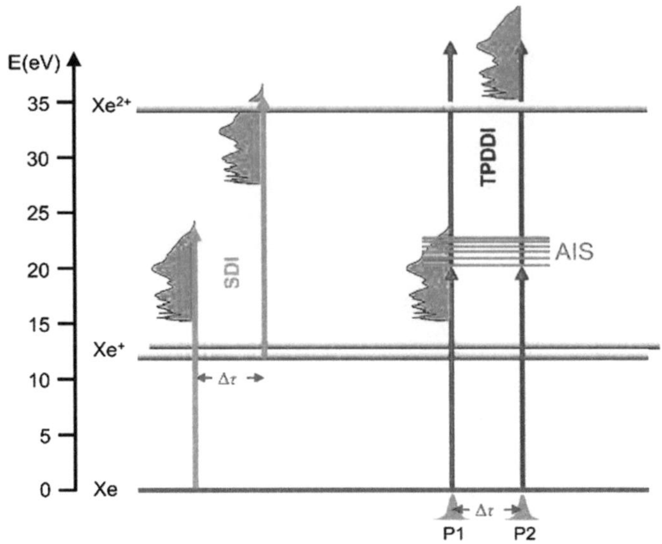

Fig. 5.10 Excitation scheme of Xe showing the main double ionization channels [30]

sequence of two XUV pulses and varying the delay $\Delta\tau$ between them, the trace of Fig. 5.11a, b was measured and will be discussed in the following sections.

5.4.2 The Second Order IVAC of the Continuum Radiation

The maximum around $\Delta\tau = 0$, where the two pulses are overlapping, is the second order IVAC of the XUV continuum. This is because contributions to this part of the trace are predominantly by the TPDDI channel (a) due to the shortness of the pulse and (b) because only a small portion of the spectrum contributes to the sequential channels. A Gaussian distribution fit (Fig. 5.11a, b) results a pulse duration of 1.5 fs. Due to the shot-to-shot CEP variation, the spectrum varies from pure to modulated continuum, i.e., in the time domain the waveform varies between a single pulse and pulses with a double peak structure. Thus the measured duration of 1.5 fs is the average of the duration of single pulses, most probably of sub-fs duration and double maxima pulses with a peak separation half the laser period (1.33 fs). Consequently, the measured duration is $\tau_{\mathrm{XUV}} = 1.5^{+0.2}_{-0.7}$ fs (800 asec being the duration for which the double peak structure is not any more resolvable in the AC trace). This duration further verifies the dominance of the TPDDI channel. An SDI trace would be the cycle average of the square of the first-order autocorrelation of the field and thus would include structures (three peaks originating from the pulses having double peak structure, the highest of which appears at zero delay) having a width equal to the width of the FT of the XUV spectrum (420 asec). The much larger measured

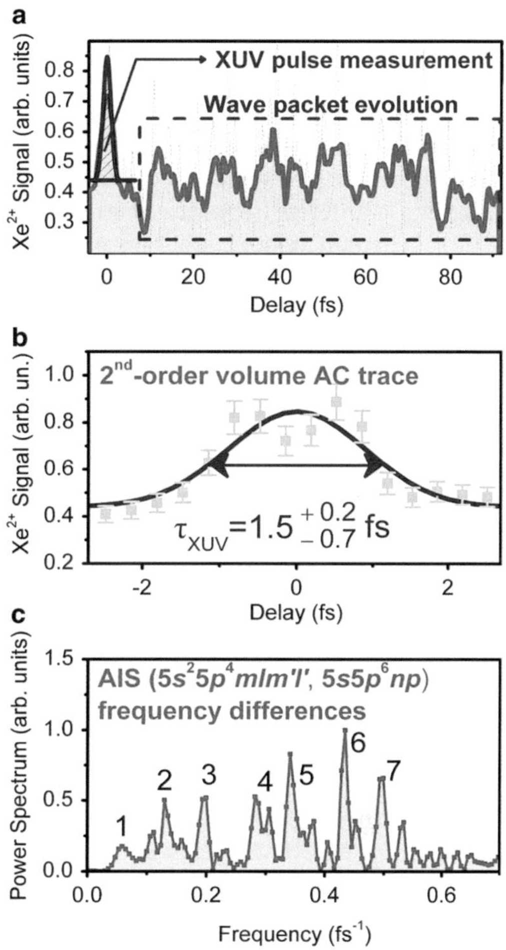

Fig. 5.11 (**a**) Second-order autocorrelation trace retrieved by recording the dependence of the Xe^{2+} signal on the delay between the XUV pulses. The *dashed line* corresponds to the raw data. The running averages over nine delay points are shown with the *solid line*-filled area). The interval around zero delay (–3–6 fs) serves for the XUV pulse characterization and the trace at longer delays (6–92 fs) measures the ultrafast evolving induced atomic coherences. (**b**) An expanded area of the trace in a. The Gaussian fit (*solid line*) on the raw data yields $\tau_{XUV} = 1.5^{+0.2}_{-0.7}$. (**c**) FT of the trace a at delays >6 fs reviling the frequency differences of the AIS [30]

width confirms that the two electrons are ejected "together" before the system finds the time to first decay in $Xe^+ + e^-$. TPDDI was observed before [70] only under conditions for which the SDI is a two-photon process [71]. Thus this is the first experiment that demonstrates observation of TPDDI under conditions at which the SDI is also a two-photon process.

5.4.3 *XUV-Pump-XUV-Probe of an Atomic Coherence*

The sequence of the two pulses is inducing and controlling a fast-evolving coherence in the structured continuum, while the second pulse at longer delays, i.e., the part of the trace where the two pulses are not overlapping evaluates this evolution, probes the evolution of the coherence. Here the first pulse (P1) induces the coherence, pumps a coherent superposition of the AIS manifold and ionic states. It also partially doubly ionizes the atom ejecting part of the formed electron wave packet, thus contributing with a background to the Xe^{2+} signal. The second pulse (P2) (1) excites a replica of the wave packet, which interferes with the evolved first one. Thus, it controls the excitation by means of constructive or destructive interferences occurring at different frequencies corresponding to the excitation frequencies of the AISs, (2) partially ionizes the wave packet it excites (background), and (3) probes the evolution of the first excited wave packet. The path involving excitation by P1 and double ionization by P2 is to the first successful experimental implementation of an XUV-pump-XUV-probe sequence at the 1-fs temporal scale.

The beating signal (Fig. 5.11a) for delays larger than 5 fs is the result of the XUV-pump-XUV-probe process. At the present temporal resolution high-frequency Ramsey fringes corresponding to the excitation frequencies of each AIS are averaged out and the low-frequency pump-probe components of the temporal evolution of the coherent superposition of the AIS dominate the trace [72]. Contributions from modulations in the excitation process due to the two interfering wave packets undergoing the same evolution are not to be excluded [73, 74]. Fourier transform (FT) of the traces reveals frequency differences of the AIS involved (Fig. 5.11c). Each frequency peak is an overlap of few not resolved spectral lines. A substantial increase of the total length of the temporal delay would allow resolving all frequency components. An assignment of the FT spectrum peaks can be found in [29], where the wavepacket dynamics were observed at the autoionization process (single-photon process) and in [30]. The correlation between the Xe^{2+} [30] and Xe^{+} [29] temporal traces is fairly visible in their FT spectra, within the frequency error of $\pm 30 \times 10^{-3}\,\mathrm{fs}^{-1}$ resulted by the temporal sampling window. Additional frequencies in the present experiment can arise from the coherence induced in the superposition of the two-fine structure levels $P_{1/2}$ and $P_{3/2}$ of the Xe^{+} ground state (beat frequency $0.31\,\mathrm{fs}^{-1}$) probed through the SDI. Since the XUV pulse duration is shorter than the time interval in which the spin–orbit occurs (pulse duration shorter than the precession period), the two-fine structure levels of Xe^{+} are coherently populated. Double peak structure of the XUV pulse, due to the CEP variation, introduces in the temporal trace components shifted by half the laser period, which only reduces the fringe contrast.

At this point it is worth discussing why the beating is observable in the measured trace. As mentioned above the high-frequency Ramsey fringes are not resolvable. In case the trace of the high-frequency Ramsey fringes would be symmetric with respect to the signal at delay $\Delta\tau \rightarrow \infty$, averaging out of the fringes would lead to signal depicting no modulation and equal to the signal at $\Delta\tau \rightarrow \infty$. From calculations performed in helium [30], it became apparent that the key element of the observed signal is its dependence on the field-dependent amplitudes of the Raman couplings between the autoionizing states through all single-photon accessible states. Due to these Raman couplings the high frequency Ramsey fringes become asymmetric. The couplings are present because the bandwidth of

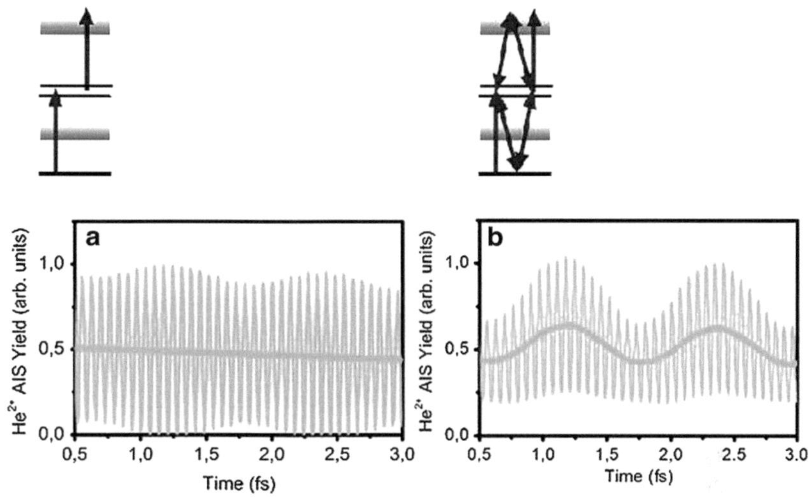

Fig. 5.12 Calculated Ramsey fringes in He in a pump-probe experiment exciting and probing the superposition of the 2s2p and 2s3p AIS. (**a**) In the calculation of the *left panel* no Raman couplings are included. The trace is symmetric and the cycle averaged signal shows no temporal dependence. (**b**) In the *right panel* Raman couplings are included in the calculation resulting to low-frequency fringing corresponding to the energy difference of the two AIS

the radiation is larger than the spacing of the AIS. They cause delay dependent redistribution of the populations of the AIS.

The population redistribution is periodic with periods equal to the inverse of the frequency distances between the states. The periodic modulation of the populations causes in turn a periodically modulated double ionization amplitude ratio of the different AIS and thus asymmetric interference fringe amplitude with respect to the double ionization yield $\Delta\tau \rightarrow \infty$. This is the reason why averaged signal shows modulation with delay. If the Raman couplings are artificially ignored in the calculation the averaged signal becomes delay independent, averaging out at all delays to its value at $\Delta\tau \rightarrow \infty$. The situation is illustrated in Fig. 5.12. It is thus due to these Raman couplings that the low-frequency beating is observable.

5.5 Conclusions and Outlook

We have presented recent systematic developments and a complete methodology allowing time domain applications and XUV pulse metrology approaches at the boundary between femto- and attosecond scales, based on two-XUV-photon processes. The tools and methods presented are directly applicable to isolated attosecond pulses. The developments include approaches for the generation of energetic ultra-broadband coherent continuum XUV radiation, through temporally confined frequency up conversion of high-peak-power, many-cycle laser pulses;

achieving conditions, at which the direct two-photon two-electron ejection channel becomes more efficient than the sequential one, permitting extension of second order autocorrelation measurements to shorter XUV wavelengths as earlier; implementation of non-linear autocorrelation of the above continuum radiation; demonstration of the necessity of second order IVAC-based approaches in particular under non a priori known conditions for which RABITT-based approaches fail; and an approach for on line shot-to shot monitoring of the absolute CEP value of the many-cycle pulses, appropriate to be used as single asec pulse piker. At the present stage, the developments discussed represent a series of first proof of principle demonstrations, achieved at the edge of the possible. In the near future, subject to the development of new advanced laser technologies and thus of advanced intense asec pulse sources, this type of experiments will improve in statistics, resolution and will be amended by more sophisticated detection schemes, allowing accurate studies of sub-fs-scale dynamics. In conjunction with the spatially confined occurrence of nonlinear processes, the developed route provides access to 4D studies at highest possible spatiotemporal resolution and is expected to become a central research instrument in the upcoming advanced attosecond research infrastructures.

Acknowledgements This work is supported in part by the European Commission programmes ULF, ALADIN (Grant Agreement No. 228334), ATTOFEL (Grant Agreement No. 238362), FASTQUAST (PITN-GA-2008-214962), ELI-PP (Grant Agreement No. 212105) and FLUX program (Contract No. PIAPP-GA-2008-218053) of the 7th FP. We thank G. Konstantinidis and A. Kostopoulos for their assistance in developing special optical components. L.A.A.N. acknowledges support from COST CM0702 action and ICHEC at Dublin.

References

1. P. Agostini L.F. DiMauro Rep. Prog. Phys **67**, 813–855 (2004)
2. P.B. Corkum, F. Krausz Nature Phys **3**, 381–387 (2007) and references therein
3. F. Krausz, M. Ivanov Rev. Mod. Phys **81**, 163 (2009)
4. G. Sansone et al., Science **314**, 442 (2006)
5. E. Goulielmakis et al. Science **320**, 1614 (2008)
6. M. Hentschel et al., Nature **414**, 509 (2002)
7. R. Kienberger et al., Nature **427**, 817 (2004)
8. A. Baltuska et al., Nature **421**, 611 (2003)
9. F. Ferrari et al., Nature Photonics **4**, 875, (2010)
10. S. Witte et al., Opt. Express **14**, 8168 (2007)
11. D. Herrmann et al., Opt. Lett. **34**, 2459 (2009)
12. D. Descamps et al., Phys. Rev. A. **64** 031404 (2001)
13. Y. Kobayashi et al. Opt. Lett. **23**, 64 (1998)
14. T. Sekikawa et al. Phys. Rev. Lett. **23**, 64 (1998)
15. T. Sekikawa et al. Phys. Rev. Lett. **88**, 193902 (2002)
16. A. Peralta Conde et al., Phys. Rev. A. **79**, 061405 (2009)
17. N.A. Papadogiannis et al., Phys. Rev. Lett. **90**, 133902 (2003)
18. P. Tzallas et al., Nature **426**, 267 (2003)
19. P. Tzallas et al., J. Mod. Opt. **52**, 321 (2005)
20. Y. Nabekawa et. al. Phys. Rev. Lett. **96**, 083901 (2006)
21. Y. Nabekawa et. al. Phys. Rev. Lett. **97**, 153904 (2006)

22. Y. Namura et al., Nature Physics **5**, 124 (2009)
23. R. Hörlein et al., New J. Phys. **12**, 043020 (2010)
24. P.M. Paul et al., Science **292**, 1689 (2001)
25. J. Kruse, et al., Phys. Rev. A. **82**, 021402(R) (2010)
26. J. Kruse, et al., Phys. Rev. A. **82**, 033438 (2010)
27. P. Tzallas, et al., Nature Phys. **3**, 846 (2007)
28. E. Skantzakis et al., Opt. Lett. **34**, 1732 (2009)
29. E. Skantzakis et al., Phys. Rev. Lett. **105**, 043902 (2010)
30. P. Tzallas et al., Nature Phys. DOI: 10.1038/NPHYS2033 (2011)
31. P. Tzallas et al., Phys. Rev. A. **82**, 061401(R) (2010)
32. E. Constant et al., Phys. Rev. Lett. **82**, 1668 (1999)
33. E.J. Takahashi et al., IEEE J. Sel. Top. Q. Elec. **10**, 1315 (2004)
34. J.L. Krause et al., Phys. Rev. A. **45**, 4998, (1992)
35. J.F. Hergott et al., Phys. Rev. A. **66**, 021801(R) (2002)
36. J. Seres et al., Nature Phys. **3**, 878 (2007)
37. F. Quere et al., Phys. Rev. Lett. **96**, 125004 (2006)
38. A. Tarasevitch et al., Phys. Rev. Lett. 98, 103902 (2007)
39. F. Quere et al., Phys. Rev. Lett. **100**, 095004 (2008)
40. C. Thaury et al., Nature Phys. **4**, 631 (2008)
41. S.V. Bulanov et al., Phys. Plasmas. **1**, 745 (1994)
42. G.D. Tsakiris et al., New J. Phys. **8**, 19 (2006)
43. T. Baeva et al., Phys. Rev. E. **74**, 046404 (2006)
44. B. Dromey et al., Nature Phys. **2**, 456 (2006)
45. S. Gordienko et al., Phys. Rev. Lett. **93**, 115002 (2004)
46. R. Lichters et al., Phys. Plasmas **3**, 3425 (1996)
47. P. Tzallas et al., *Progress in Ultrafast Intense Laser Science*, vol. VII. (Springer, New York) p.163, (2010) (ISBN: 978-3-642-18326-3)
48. O. Tcherbakoff et al., Phys. Rev. A.**68**, 043804 (2003)
49. C. Altucci et al., Opt. Lett. **33**, 2943 (2008)
50. H. Mashiko et al., Phys. Rev. Lett. **100**, 103906 (2008)
51. E. Takahashi et al., Phys. Rev. Lett. **104**, 233901 (2010)
52. V.V. Strelkov et al., New J. Phys. **10**, 083040 (2008)
53. I.J. Sola et al.,Nature Phys. **2**, 319 (2006)
54. G. Sansone, Phys. Rev. A. **79**, 053410 (2009)
55. C. Iaconis, I.A. Walmsley Opt. Lett. **23**, 792 (1998)
56. M. Drescher et al., Science**291**, 1923 (2001)
57. Y. Mairesse, F. Quere, Phys. Rev. A. **71**, 011401(R) (2005)
58. E. Cormier et al., Phys. Rev. Lett. **94**, 033905 (2005)
59. Y. Mairesse et al., Phys. Rev. Lett. **94**, 173903 (2005)
60. N. Dudovich et al., Nature Phys. **2**, 781 (2006)
61. J.M. Dahlstrom et al., Phys. Rev. A. **80**, 033836 (2009)
62. P.B. Corkum Phys. Rev. Lett. **71**, 1994 (1993)
63. M. Lewenstein et al., Phys. Rev. A. **49**, 2117 (1994)
64. Ph. Antoine et al., Phys. Rev. Lett. **77**, 12347 (1996)
65. M.B. Gaarde et al., Phys. Rev. Lett. **89**, 213901 (2002)
66. P. Salieres et al., Phys. Rev. Lett. 74, 3776 (1995)
67. M. Bellini et al., Phys. Rev. Lett. 81, 297 (1998)
68. P. Lambropoulos et al., Phys. Rev. A. **78**, 055402 (2008)
69. T. Nakajima, L.A.A. Nikolopoulos Phys. Rev. A. **66**, 041402(R) (2002)
70. E. Benis et al., Phys. Rev. A. **74,** 051402 (2006)
71. M.A. Kornberg, P. Lambropoulos J. Phys. B. **32**, L63 (1999)
72. V. Blanchet et al., J. Chem. Phys. **108**, 4862 (1998)
73. B. Campbell et al., Phys. Rev. A. **57**, 4616 (1998)
74. R.R. Jones et al., Phys. Rev. Lett. **71**, 2575 (1993)

Chapter 6
Controlling the Motion of Electronic Wavepackets Using Cycle-Sculpted Two-Color Laser Fields

M. Kitzler, X. Xie, S. Roither, D. Kartashov, and A. Baltuška

Abstract We use cycle-sculpted two-color waveforms to drive electronic wavepackets generated by strong-field ionization from helium, neon, and argon gas atoms and analyze their momentum spectra measured by electron-ion coincidence momentum spectroscopy. Varying the relative phase of the two colors allows to sculpt the ionizing field and hence to control the emission times and motion of the wavepackets on an attosecond timescale. Using semiclassical calculations, we investigate the influence of the ionic Coulomb field onto the motion of emitted electronic wavepackets. We further show that the measured electron momentum spectra contain interference patterns created by pairs of electron wavepackets that are released within a single laser-field cycle. We experimentally distinguish these subcycle interference structures from above-threshold ionization (ATI) peaks and argue that they can be used to extract the subcycle phase evolution of the laser-driven complex bound-state wavefunction.

6.1 Introduction

Ultrashort intense laser pulses are a unique tool to create coherent electron wavepackets with subcycle duration via the strongly nonlinear process of tunnel ionization of atoms or molecules [1, 2]. After their creation, the wavepackets are driven by the combined forces of the laser electric field and the ionic binding potential [3]. It is thus possible to actively control their motion in space [4, 5] and time by controlling the laser electric field [2]. When the wavepackets are directed

---

M. Kitzler (✉) · X. Xie · S. Roither · D. Kartashov · A. Baltuška
Photonics Institute, Vienna University of Technology, Gusshausstrasse 27/387, 1040 Vienna, Austria
e-mail: markus.kitzler@tuwien.ac.at

K. Yamanouchi et al. (eds.), *Progress in Ultrafast Intense Laser Science VIII*,
Springer Series in Chemical Physics 103, DOI 10.1007/978-3-642-28726-8_6,
© Springer-Verlag Berlin Heidelberg 2012

back to the ion core, they can be used to coherently probe the parent ion on the attosecond and Ångström scale by recording suitable probe signals such as electron momenta or energies, e.g., [6–10], or spectra of photons emitted via the process of high-harmonic generation, e.g., [11–15].

Recording the momentum spectra of coherently generated electron wavepackets allows for interferometric measurements: When the spectra overlap at their time of detection, one observes interference structures [16–20] very similar to the interference fringes of coherent optical pulses imprinted onto optical intensity spectra [21, 22]. A well-known type of interference structures created by electron wavepackets is the above-threshold ionization (ATI) peaks in photoelectron energy spectra [23]. These are generated by wavepackets released by strong-field ionization during different cycles of period T of an intense laser field [24] and are spaced by the laser photon energy $\hbar\omega$ with $\omega = 2\pi/T$ the laser frequency. Interfering pairs of wavepackets, however, can also be created within the same laser cycle [7, 24–26] released with delays smaller than T. They lead to interference fringes of subcycle origin which are visible as a modulation of the ATI-peak structures. Their relative subcycle release time Δt is encoded in momentum with smaller Δt interfering at higher momenta.

When pairs of wavepackets are launched at different instants of time, their differences in quantum phases are accumulated during ionization and propagated in the presence of both the ionic and laser field. It is thus possible to actively control and modify the interference fringes with cycle-sculpted laser fields. In this chapter, we describe experiments that exploit this opportunity for helium atoms using a two-color incarnation [27, 28] of such laser fields.

Using two-color sculpted laser fields not only allows to control the accumulated quantum phases that determine the interference structures, but also the attosecond temporal evolution of the driving laser force of the wavepackets that create them. The wavepackets' motion after ionization is, however, not only determined by the laser pulse but by the combined force of the laser-electric field and the ionic Coulomb potential. Interpretation of experimental electron or photon spectra is, nevertheless, usually performed within the framework of the strong-field approximation (SFA) [29] that neglects the influence of the Coulomb potential on electron trajectories, which might lead to severe errors. The relevance of the Coulomb contribution has been demonstrated in many numerical simulations, e.g., [30–33]. Inclusion of the Coulomb force into the theoretical description of, e.g., electron momentum spectra is not straightforward, though [34].

Unfortunately, it is in general not possible to separately measure only the contributions of the Coulomb potential to the probe signal and to quantify their influence experimentally, which would allow to estimate the errors made in the SFA. In our experiments, we use sculpted laser fields. Therefore, we can control the driving force of the laser field on an attosecond timescale. By measuring electron momentum spectra for a range of field shapes, we can extract the influence of the Coulomb potential by comparing measured spectra to semiclassical and quantum mechanical models. Our experiments show a strong influence of the Coulomb field on the final momentum of field-ionizing electron wavepackets. We are able to trace back its influence to the timing of the wavepacket release within the laser field cycle.

This chapter is organized as follows: First, we describe the experimental setup. Then we present measured electron-ion momentum spectra. We show that they contain signatures of a pronounced influence of the Coulomb potential onto the motion of driven wavepackets and discuss the underlying mechanisms of the Coulomb distortion. After that, we demonstrate that the measured raw spectra also contain subtle signatures of electron wavepacket interferences which will be analyzed in the remainder of this chapter. It will be shown that the interference patterns encode the quantum phase difference of pairs of electron wavepackets that are timed with attosecond precision.

6.2 Experiment

In our experiments, we generated cycle-sculpted field cycles by coherently superimposing a strong ~30 fs (full width at half maximum, FWHM) 790-nm laser pulse, frequency ω, and its second harmonic, frequency 2ω, generated in a type-I BBO crystal (thickness 500 μm), with parallel polarization directions in a collinear inline geometry, such that the total electric field of the two-color pulse can be written as

$$E(t) = \hat{E}[f_1(t)\cos(\omega t) + f_2(t)\cos(2\omega t + \varphi)], \tag{6.1}$$

with $f_i(t)$, $i = 1, 2$, Gaussian pulse envelopes with maximum values of 1, φ the two colors' relative phase, and $\hat{E}$ the peak electric field which is related to the pulse peak intensity as $\hat{E} = \sqrt{I}$. Here and throughout this paper atomic units (au) are used unless otherwise stated. The duration of the 2ω pulse was determined from cross-correlation measurements using the 3ω signal to be roughly twice as long as the fundamental pulse. The relative pulse energy was adjusted by varying the angle of reflection off a glass plate utilizing the different reflection coefficients for p- and s-polarized light, such that in the focus of the beams, the field strengths were equal, taking into account the slightly different beam diameters and divergences as well as the tighter focusing of the 2ω pulse. The polarization directions of the two pulses were set parallel to each other by means of a $\lambda/2$ waveplate for $\lambda = 790$ nm. The pulse repetition rate was 5 kHz. The temporal overlap of the two pulses was ensured by compensating for the two colors' different group velocities with calcite plates, and the relative phase of the two pulses was varied by introducing one wedge of a pair of fused silica wedges into the combined red and blue laser beam. Similar two-color fields in this so-called $\omega/2\omega$ configuration have been experimentally applied in the investigation of above-threshold ionization; see, e.g., [27, 28, 35] and references therein, and in controlling ionization and fragmentation [36–38] as well as orientation [39] of molecules. The maximum cycle asymmetry of the pulses (achieved for $\varphi = 0$) is equivalent to the one of a true single-cycle pulse of one

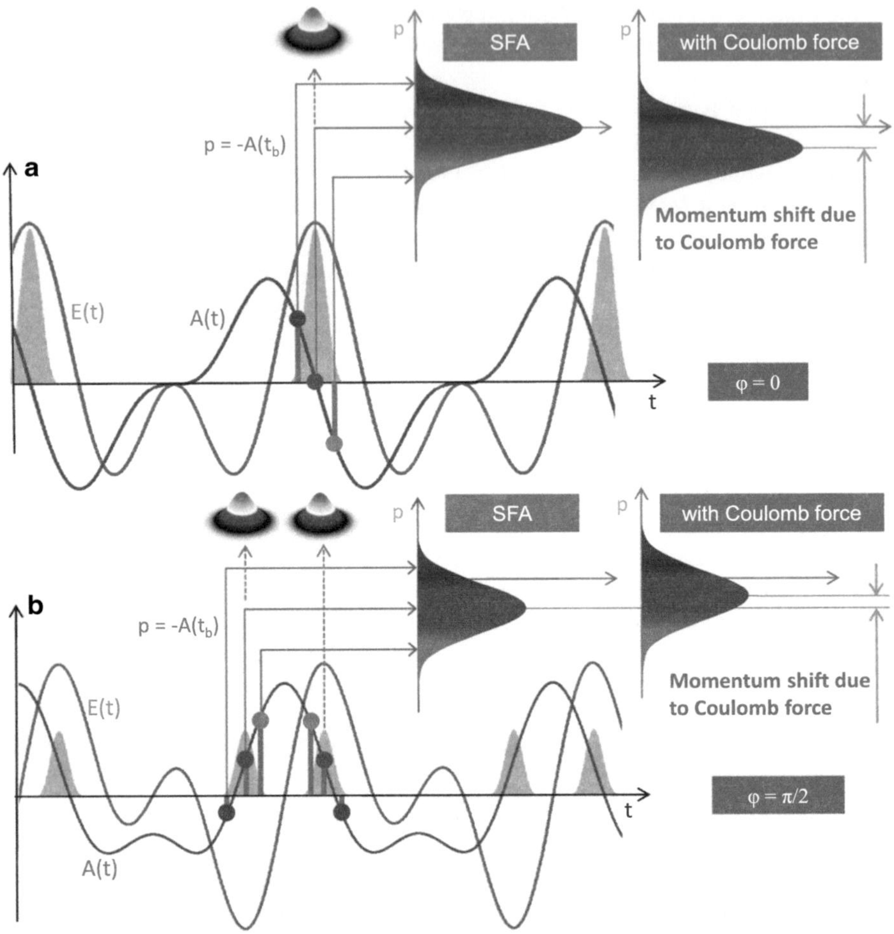

Fig. 6.1 Ionization and wavepacket motion with sculpted two-color laser pulses. Electron wavepackets are emitted at the crests of the electric field $E(t)$ of a laser pulse. Without the influence of the Coulomb potential, an electron born at some time t_b reaches a final momentum given by the negative vector potential at birth time, $p = -A(t_b)$. By varying φ, the timing of the wavepackets' creation as well as their motion after ionization can be controlled. (**a**) The shape of the field cycle for $\varphi = 0$ is strongly asymmetric with one strong peak per laser cycle. Because the shape of the vector potential sweeps through zero during wavepacket emission, without the Coulomb potential's force, the observed electron momentum spectrum is expected to be centered around zero. The Coulomb potential influences the motion of the wavepackets and leads to a distortion of the spectral shape and a shift of the spectral mean value. (**b**) For $\varphi = \pi/2$, two wavepackets per cycle are emitted at times when $A(t)$ is positive. Therefore, without the Coulomb potential, the spectral mean value is shifted towards negative values. The Coulomb force leads to a decreased shift and a distortion of the spectral shape

color with a controlled carrier-envelope phase, which for 790 nm is only achieved for pulse durations of 2.6 fs. Varying the relative phase φ of the two colors allowed us to sculpt the ionizing field and hence to control the emission times and motion of the interfering wavepackets on an attosecond timescale; see Fig. 6.1.

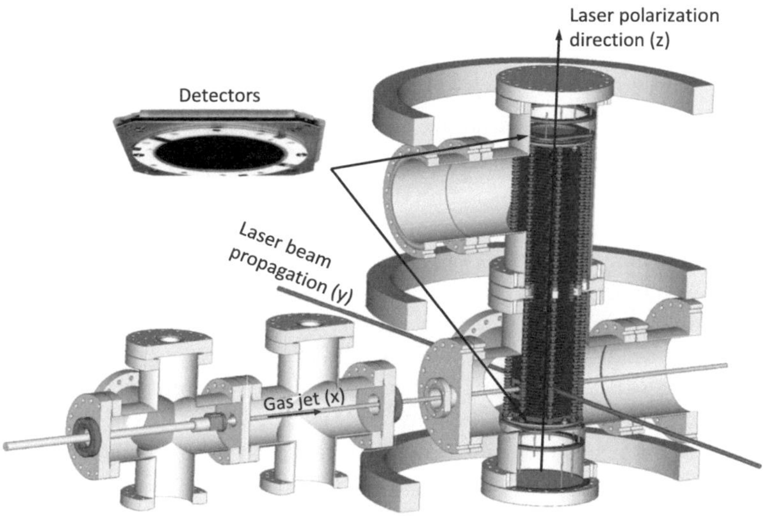

Fig. 6.2 Detection apparatus used for the measurement of electron-ion momentum spectra. Created ions and electrons are guided by weak magnetic and electric fields to two RoentDek DLD 80 detectors situated at opposite ends of the vacuum chamber. The homogeneous electric field pointing into the z-direction is produced by equidistant copper rings (inside the vacuum chamber); the weak homogeneous magnetic field (also along the z-direction) is produced by three copper coils surrounding the vacuum chamber

We use Cold Target Recoil Ion Momentum Spectroscopy (COLTRIMS) [40] to measure the three-dimensional vector of electrons and ions emerging from the interaction of a single atom with the sculpted laser fields. Our detection apparatus, see Fig. 6.2, consists of a two-stage arrangement to provide an internally cold gas jet of atoms traveling at nearly identical ultrasonic velocities and an ultrahigh vacuum chamber ($\approx 10^{-10}$ mbar) where the gas jet is crossed with the laser beam. The ions and electrons created during single ionization are guided by weak magnetic and electric fields to two RoentDek DLD 80 detectors situated at opposite ends of the vacuum chamber. The homogeneous electric field of 2.5 V/cm pointing into the z-direction is produced by equidistant copper rings; the weak homogeneous magnetic field of 6.4 G (also along the z-direction) is produced by three copper coils. With these directions of the fields, the ions are accelerated over a distance of 45 cm before they reach the upper detector, and the electrons are accelerated over 5.7 cm before reaching the lower detector. For the data analysis, we identify the ion momentum to be the mirror image of the electron momentum in all three spatial directions due to the negligible momentum transfer by the photon field [41]. We estimate the momentum resolution of our experiment to be ≈ 0.05 au along the z-direction for electrons and ions and ≈ 0.05 au along the x/y-directions for electrons.

6.3 Measured Electron Momentum Spectra and Influence of the Coulomb Field

Figure 6.3a,b shows the measured ion momentum vector component parallel to the laser field's polarization direction (z), $p_{\|}$, as a function of φ for He^+ (a) and Ar^+ (b) ions, where the other momentum directions, p_x and p_y, have been integrated over. Due to the negligible momentum transfer by the photon field, the momentum of the ion is the mirror image of the electron momentum [41]. The pulse peak intensity for each of the two colors was 1×10^{14} W/cm^2 for helium and 2×10^{13} W/cm^2 for argon. The mean value of the measured spectra for helium is shown by the light gray line in Fig. 6.3a. It strongly varies with φ and shows extrema for $\varphi \approx (n + 0.3)\pi$, $n \in \mathbb{Z}$. The spectral intensity, encoded in gray scale, also shows a pronounced dependence on φ and has maxima for $\varphi = n\pi$ and minima for $\varphi = (n + 1/2)\pi$, $n \in \mathbb{Z}$. The dependence of both the mean value and the intensity on φ is stronger for helium than for argon. In the following, we will model the observed spectra and their dependence on φ using semiclassical methods and investigate the influence of the Coulomb field on the motion of the field-driven electron wavepackets. As a first step, we model the spectra assuming a purely field-driven motion of the electronic wavepackets after ionization, in accordance with the SFA.

6.3.1 *Field-Driven Wavepacket Motion*

Within the SFA, the momentum spectrum observed along the field's polarization direction after the laser pulse has faded is given by integration of the relation $p_{\|} = -A(t_b)$, with $A(t) = -\mathrm{d}E(t)/\mathrm{d}t$ being the vector potential, over all possible birth times t_b, at which the wavepacket is emitted with a probability determined by the ionization rate, e.g., given by [42]. Figure 6.1 visualizes the birth time to momentum mapping for the two field shapes generated by adjusting φ to 0 (a) and $\pi/2$ (b), respectively. The ionization rate, i.e., the production rate of electron wavepackets, is determined by the electric field $E(t)$, but the final momentum of an electron is determined by the vector potential $A(t)$. As shown in Fig. 6.1, the electric field $E(t)$ for $\varphi = 0$ is strongly asymmetric with one strong positive peak per laser cycle, but the vector potential $A(t)$ has a symmetric shape. Due to the strongly nonlinear ionization rate, wavepacket emission takes place mostly around the positive peaks of $E(t)$ (gray area in Fig. 6.1). The vector potential $A(t)$ around these times passes through zero. Therefore, within the SFA, the electron momentum distribution is predicted to be symmetrically centered around zero. For $\varphi = \pi/2$, the electric field $E(t)$ has a symmetric shape. Wavepacket emission takes place twice per laser cycle around the extrema of $E(t)$. Because of the asymmetric shape of the vector potential $A(t)$ that shows positive peaks at the times of emission, the SFA predicts a final electron momentum distribution centered around a pronounced negative offset (see Fig. 6.1). We show electron momentum spectra for $p_{\|}$ for a

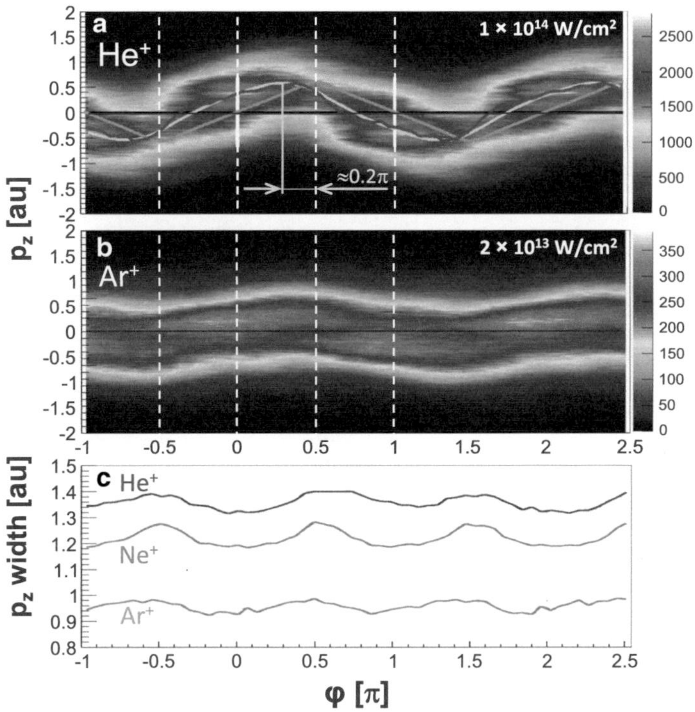

Fig. 6.3 Momentum component parallel to the laser polarization direction, $p_{\parallel}$, of ions created during single ionization of He (**a**) and Ar (**b**), as well as (**c**) the width of the momentum spectra for He, Ne, and Ar as a function of the relative phase, φ, between the ω and 2ω components of the two-color laser field. The mean momentum value of the measured data in (**a**) is shown by the light line that peaks at $\varphi \approx 0.3\pi$. Predictions of a classical model that only considers the driving force of the laser field, but neglects the influence of the Coulomb potential, peak at $\varphi = 0.5\pi$. When the Coulomb potential is included into the model, the simulated spectral mean value agrees well with the measured mean value (*dark line*)

range of relative phases calculated by this SFA model in Fig. 6.4a. As expected, their mean value shows a strong modulation with φ, with maxima and minima for $\varphi = (n + 1/2)\pi$ and zero mean value for $\varphi = n\pi$, $n \in \mathbb{Z}$. To obtain these spectra, we approximated the emitted wavepackets by an ensemble of classical trajectories, each of them resembling an emitted point charge. The trajectories are launched at every time grid step used in the calculation and are weighted by the ionization rate at that time. After the laser pulse, the weighted momentum spectrum of all trajectories is calculated from their final momentum given by $p_{\parallel} = -A(t_b)$, with t_b the time when they are launched. This classical model, except for effects that are caused by wavepacket interferences, which it does not reproduce, is equivalent to the so-called semiclassical simple man's model [25] and serves as our SFA reference model. In addition to the variation of the spectral mean value with φ, Fig. 6.4a shows a similar modulation of the spectral intensity (encoded in gray scale) as the experimentally

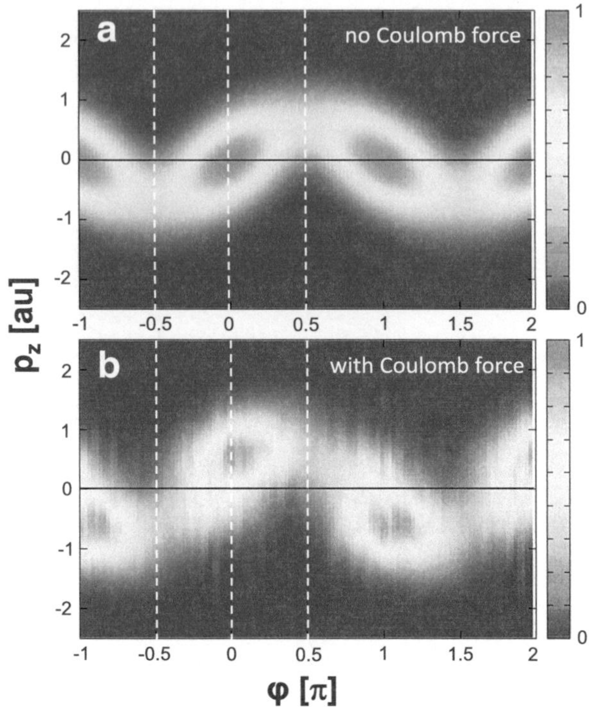

Fig. 6.4 Momentum component parallel to the laser polarization direction, $p_{\parallel}$, of He^{+} simulated using a semiclassical model for purely field-driven electrons (**a**) and with the Coulomb potential included (**b**) as a function of the relative phase, φ, between the ω and 2ω components of the two-color laser field

observed spectra. This is because the crests of the electric field $E(t)$ are smaller for $\varphi = \pi/2$ than for $\varphi = 0$. Because the ionization rate is strongly nonlinear in $E(t)$, the ionization yield per pulse and therewith the spectral intensity is larger for $\varphi = 0$ than for $\varphi = \pi/2$. This explains the intensity modulation of the spectra that is observed experimentally and also predicted by the SFA.

6.3.2 Influence of the Coulomb Field on the Spectral Mean Value

We show the spectral mean value of the calculated spectra shown in Fig. 6.4a as a function of φ by the gray line that peaks at $\varphi = (n + 1/2)\pi$, $n \in \mathbb{Z}$, in Fig. 6.3a. While the experimentally observed strong modulation with φ is in accordance with the SFA model, the phases at which the mean momenta reach extrema do not coincide with those of the SFA prediction. A phase shift between the measured and SFA maxima of about 0.2π (Fig. 6.3a) is observed which results in a large spectral offset for $\varphi = 0$, for which the SFA model predicts zero momentum. It is thus incorrect to calibrate the field cycle shape using the maxima of the spectral asymmetry, as is commonly done, e.g., in [26]. We will now demonstrate that the pronounced deviations between the model predictions and the measurements are due to the influence of the Coulomb potential. To that end, we incorporate the long-range Coulomb potential into the above described trajectory model. As before, we approximate a wavepacket by an ensemble of classical trajectories launched at time t_b and weighted by the ionization rate at that time. This time, however, the point charges are not only driven by the laser electric field but also by the force due to the Coulomb potential which we assume in the form of a soft-core potential $V(z) = -1/\sqrt{z^2 + a}$ with $a = 1$. The classical trajectories are started with zero initial velocity at the tunneling birth point, $z(t) = -I_p/E(t)$, which is widely independent of a (here I_p is the ionization potential of the atom), and are propagated numerically. After the laser pulse, the electron momentum spectrum is obtained by calculating the weighted momentum spectrum of all trajectories. The spectra obtained by this model are shown in Fig. 6.4b as a function of φ. Indeed we observe the maxima and minima of the mean momentum at different positions than with the SFA model, and the spectrum for $\varphi = 0$ shows a pronounced mean momentum different from zero. The overall appearance much resembles the experimentally observed spectra in Fig. 6.3a. For a quantitative comparison with the experimental spectra, we calculate the mean momenta from the numerically obtained spectra and plot them by the dark gray line in Fig. 6.3a. The measured shift of the spectral mean value for $\varphi = 0$ as well as the shift of the maximum value of 0.2π from the SFA value is exactly reproduced. We repeated the same calculations by solving the three-dimensional time-dependent Schrödinger equation (TDSE) for a single active electron in a model potential of helium [43]. These simulations reproduce the mean momentum value calculated by the semiclassical model almost exactly.

6.3.3 *Influence of the Coulomb Field on the Spectral Width*

We have shown that due to the influence of the ionic Coulomb field, the mean value of observed electron spectra is strongly shifted as compared to purely field-driven electron wavepackets. We will now show that not only the spectra's mean value but also their shape, in particular their width, is strongly affected by the Coulomb potential. The width of the spectra in $p_{\parallel}$ calculated from the measured spectra for He, Ne, and Ar is shown in Fig. 6.3c. The data for Ne has been taken with the same pulse as the one for He. The spectral width shows maxima for symmetric electric field shapes at $\varphi = \pi/2$ and minima for asymmetric electric field shapes at $\varphi = 0$. Because the symmetric field exhibits smaller peak electric fields, this seemingly contradicts tunneling theory which predicts the opposite: broader momentum distributions for higher electric fields [44].

The contradiction to tunneling theory can be resolved if we consider the shape of the spectra for a certain φ. In the left column of Fig. 6.5, we show measured spectra for $\varphi = 0$ and $\varphi = \pi/2$ for helium and argon, respectively. The larger width of those spectra taken at $\varphi = \pi/2$ as compared to those taken at $\varphi = 0$ can be clearly seen. In the spectrum for argon at $\varphi = \pi/2$, a prominent dip in the center of the

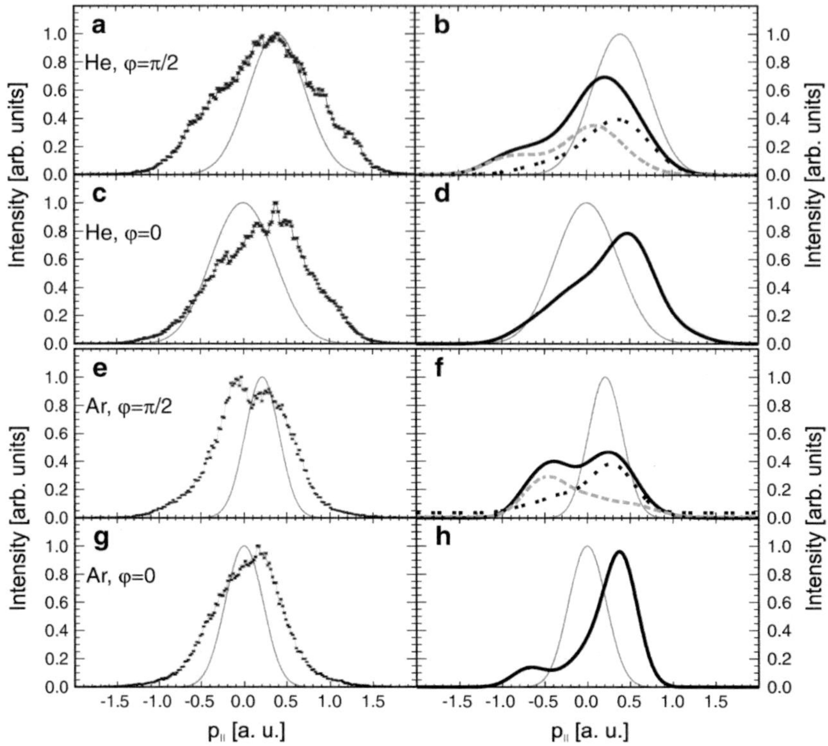

Fig. 6.5 Measured (*left column*) and simulated (*right column*) electron momentum spectra for He (**a**)–(**d**) and Ar (**e**)–(**h**) in the direction parallel to the laser polarization direction for the two-color relative phases $\varphi = 0$ and $\varphi = \pi/2$. The experimental data is shown with statistical error bars. SFA predictions are shown by *full light lines*. Model predictions with a soft-core potential included for wavepackets emitted during the positive (*negative*) half-cycle are shown by the *light dashed (dark dotted) lines*. The *full dark line* shows the sum of both

spectrum shows up. None of these features are reproduced by the SFA model (full light lines). Obviously it is thus not the laser field, but the Coulomb potential that causes the spectral distortions and the variation of the shape and width of the spectra with φ. In order to understand the origin of the spectral distortions, we compare the measured spectra to spectra simulated using the model explained above with the Coulomb potential $V(z)$ included.

First, we concentrate on the simulated spectra for helium and $\varphi = \pi/2$ shown in Fig. 6.5b. For $\varphi = \pi/2$, the electric field cycle consists of two equally strong main peaks that each cause the emission of a wavepacket; see Fig. 6.1b. One wavepacket is emitted during the negative and one during the positive half-cycle. In the classical simulations, we can decompose the spectrum (dark full line) into the two contributions of the wavepackets emitted during the positive half-cycle (light dashed line) and negative half-cycle (dark dotted line), respectively. Because for both emission events the vector potential sweeps through the same range of positive values, cf. Fig. 6.1b, within the SFA, the resulting momentum spectra should be identical for both wavepackets and centered at the same positive value, as shown by the full light line in Fig. 6.5b. With the Coulomb potential included, the results of the simulation show that the wavepacket born during the negative half-cycle is not severely affected by the Coulomb potential and centers closely to the SFA spectrum. The trajectory of the wavepacket born during the positive half-cycle, however, is obviously strongly distorted by the Coulomb potential such that the final momentum spectrum centers at a different value. For helium at the intensity of 1×10^{14} W/cm^2 for each of the two colors, which results in a peak intensity of 3×10^{14} W/cm^2 for the combined field, the deviation is moderate. Its effect is that the sum of the contributions from the two wavepackets leads to a broadening of the overall spectrum, which explains the experimentally observed larger spectral width at $\varphi = \pi/2$ shown in Fig. 6.3c. For the measurements on argon, a much lower laser intensity (combined peak intensity 0.6×10^{14} W/cm^2) was used. For that low intensity, the Coulomb force relative to the laser field's driving force becomes much stronger than for the high-intensity case of helium. The classical model predicts that the trajectory of the wavepacket born during the positive half-cycle is severely distorted, which not only leads to a broadening of the overall spectrum, as observed for helium, but even to a splitting of the spectral shape in Fig. 6.5f—exactly as observed in the measurement; cf. Fig. 6.5e. The model, however, is too simple for quantitative predictions and overestimates the energy separation of the two peaks. While a classical description of wavepacket motion is known to reproduce experimental observations well for strongly driven electrons in high electric fields [45, 46], for low field strengths, the way the ionization step is incorporated into our model might lead to artifacts. A more precise description that considers the multiphoton character of ionization, the momentum width of the tunneling wavepacket, and the uncertainty in the location of birth would be needed for quantitative statements.

Now we examine the case of the strongest field asymmetry, $\varphi = 0$. In this case, only one wavepacket per cycle is launched around an instant where the vector potential goes through zero; see Fig. 6.1a. Thus, without Coulomb potential, the momentum spectrum should be centered around zero; see spectra plotted by the light full lines in Fig. 6.5c and g. In the experiment we observe a considerable shift of the spectrum to positive ion momenta for both the helium and argon measurement. Using our simple classical model, we investigated the dynamics of the driven wavepacket that lead to the Coulomb-induced spectral shift. Our findings show that parts of the wavepacket recollide with the ion core with a very low energy and their trajectories are thus strongly affected by the long-range Coulomb potential, such that they finally reach the detector with a momentum much different from zero. This explains the nonzero mean momentum value $p_{\|}$ for $\varphi = 0$ observed in our experiments and in numerical simulations [32, 33], where a very similar Coulomb-induced shift has been predicted for carrier-envelope phase (CEP) stabilized few-cycle pulses with a spectrum centered around 800 nm.

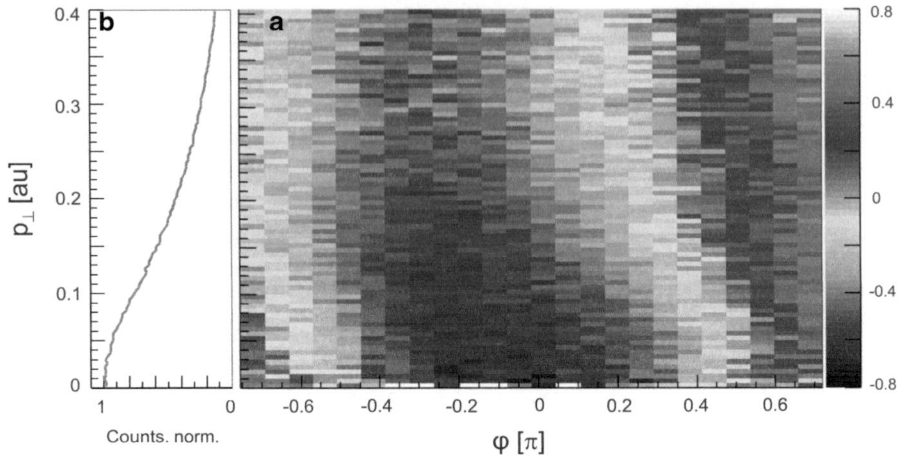

Fig. 6.6 (**a**) Asymmetry parameter derived from the measured data for helium for the momentum component parallel to the laser polarization direction, $p_\parallel$, as a function of the momentum component orthogonal to it, $p_\perp$, and the two-color relative phase φ. (**b**) Total electron yield as a function of $p_\perp$ normalized to unity at $p_\perp = 0$

6.3.4 Wavepacket Motion in the Lateral Direction

We have shown above that the force due to the Coulomb potential has a strong influence on the trajectories of a laser-driven wavepacket such that its final momentum along the laser polarization direction, $p_\parallel$, deviates considerably from that of a purely field-driven wavepacket. In our experiment, we measure not only $p_\parallel$ but also the momentum component orthogonal to the laser field polarization axis, $p_\perp$. Due to $p_\perp$, a wavepacket will laterally drift away from the ion core during its motion out of the Coulomb long-range potential towards the detector. Figure 6.6 shows the influence of the orthogonal momentum component $p_\perp$ on the shape and position of electron momentum spectra for helium observed along $p_\parallel$ for different shapes of the field cycle as determined by φ. We use the asymmetry parameter $A_\parallel = (P_+ - P_-)/(P_+ + P_-)$, with $P_\pm$ the number of electrons with positive and negative momentum value $p_\parallel$, respectively, to characterize the mean value and shape of the momentum spectra along $p_\parallel$. Figure 6.6a shows $A_\parallel$ derived from the measured spectra in Fig. 6.3a. The asymmetry parameter $A_\parallel$ serves as a measure of the deviation of the electron trajectories from those of a free electron, for which $A_\parallel$ is independent of $p_\perp$ with maxima at $\varphi = (n + 1/2)\pi$ and zero values for $\varphi = n\pi$, $n \in \mathbb{Z}$. The experimentally obtained map $A_\parallel(\varphi, p_\perp)$, however, shows a strong dependence on $p_\perp$, especially in the range $\varphi = 0 \ldots 0.5$. To explain the dependence, we plot the mean value of the electron momentum spectra along the polarization direction derived from the experimental data, $\langle p_\parallel \rangle$, for selected regions of the orthogonal momentum $p_\perp$ in Fig. 6.7. For each of the shown data points, the

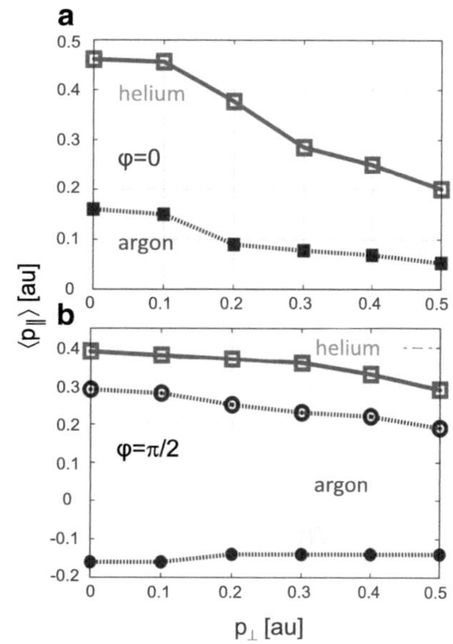

Fig. 6.7 Mean value of $p_{\|}$ for helium and argon as a function of $p_{\perp}$ for $\varphi = 0$ (**a**) and $\varphi = \pi/2$ (**b**). Open and filled circles correspond to the low and high momentum spectral peaks for argon, respectively; see Fig. 6.5e. For each of the shown data points, the spectral mean value $\langle p_{\|} \rangle$ has been calculated when the orthogonal momentum has been restricted to a range of 0.1 au centered around $p_{\perp}$

spectral mean value $\langle p_{\|} \rangle$ has been calculated when the orthogonal momentum has been restricted to a range of 0.1 au centered around $p_{\perp}$.

We first discuss the case $\varphi = 0$, Fig. 6.7a. When a wavepacket is emitted by field ionization during one half-cycle of the laser oscillation period, the wavepacket will leave the interaction zone with the Coulomb potential in different portions, depending on its emission time within the half-cycle. For a sculpted pulse with $\varphi = 0$, only one wavepacket per cycle is emitted; see Fig. 6.1a. Without the force of the Coulomb potential (as in the SFA), the fraction of the electron wavepacket that is created while the field strength increases departs directly from the parent ion on a trajectory that is commonly called a direct one. The fraction of the wavepacket that is born while the field decreases is driven back to the ion by the electric field on a recolliding trajectory [3]. Both types of electrons are first driven away from the core of the Coulomb potential and will thus experience approximately the same Coulomb action during this stage of their motion. The recolliding portion of the wavepacket, however, during its return to the parent ion, will experience an additional Coulomb force. For nonnegligible $p_{\perp}$, it will miss the center of the binding potential and pass the ion at some distance that increases with $p_{\perp}$ [47, 48]. Gating on large values of $p_{\perp}$, thus, reduces the influence of the Coulomb potential especially for those parts of the wavepacket which revisit the ion core.

This is demonstrated in Fig. 6.7a. For both helium and argon, we find that the purely Coulomb-induced large value of $p_{\|}$ at small $p_{\perp}$ gradually decreases as the influence of the recolliding electrons is reduced with increasing $p_{\perp}$. The SFA momentum value of 0 is, however, never reached, rendering, not unexpectedly, the SFA inapplicable for the interpretation of measured spectra composed of the low-energy region. The large offset from the SFA-predicted momentum value of 0 for small $p_\perp$ can be understood by considering that the wavepacket's driving laser field force is small for the asymmetric field shape $\varphi = 0$. The situation is very similar for single-cycle pulses of only one color with a carrier-envelope phase of zero. Our experiments thus show that the large Coulomb-induced spectral shift found in [32, 33] can be traced back to a strong distortion of the recolliding electron wavepackets.

Now we turn to the case $\varphi = \pi/2$ shown in Fig. 6.7b, where the shape of the electric field cycle is symmetric and two wavepackets per cycle are emitted; see Fig. 6.1b. First, we concentrate on the helium data (open squares). The mean momentum value predicted by the SFA is $+0.41$ au for both wavepackets. We can see that the measured $p_\parallel$ at $p_\perp = 0$ is close to this value. The results of the classical simulations in Fig. 6.5b show a stronger influence of the Coulomb potential on the wavepacket that is emitted during the positive half-cycle of the electric field and which is not driven back to the ion core. Thus, the small deviation is most likely caused by the contribution of the wavepacket emitted during the positive half-cycle of the electric field. As argued above, by gating on large values of $p_\perp$, we reduce the Coulomb contribution due to the recolliding wavepacket emitted during the negative half-cycle and therewith relatively enhance the contribution of the wavepacket emitted during the positive half-cycle. Consequently, $p_\parallel$ moves to smaller values, and the deviation from the SFA value becomes larger; see Fig. 6.7b.

For argon, the situation is complicated by the appearance of two distinct peaks in the momentum spectrum for $\varphi = \pi/2$. According to the classical simulations, they are caused by the same Coulomb-induced effect that for helium merely leads to a broadening of the spectrum, namely, that the wavepackets emitted during the positive and negative laser half-cycle, respectively, are effected differently by the Coulomb force. But due to the lower laser intensity used in the experiments for argon, the relative influence of the Coulomb force is stronger and the momentum splitting larger. We show the center position of the two peaks as a function of $p_\perp$ by the circles in Fig. 6.7b. For the lower momentum peak (filled circles) that is predicted to be caused by a wavepacket emitted during the negative half-cycle, $p_\parallel$ stays almost constant around -0.17 au, independent of $p_\perp$, while the value calculated by the SFA should be $+0.18$ au. For the higher momentum peak (open circles), which, according to the classical simulations, corresponds to a wavepacket emitted during the positive half-cycle, the measured $p_\parallel$ is larger than the SFA momentum value for small $p_\perp$. By gating on larger values of $p_\perp$, by which we dominantly reduce the effects of the Coulomb potential on the recolliding wavepacket and preserve mainly the effects on the wavepacket's outgoing motion away from the ion, we achieve good agreement with the SFA; see Fig. 6.7b.

In conclusion, we measured the influence of the Coulomb potential on the motion of field-ionized laser-driven electronic wavepackets by analyzing three-dimensional electron-ion coincidence momentum spectra obtained using the COLTRIMS technique. To create and steer the electron wavepackets, we used two-color laser pulses with a sculpted field cycle which allowed us to control the temporal structure of wavepacket emission and the evolution of the field's driving force on a subcycle scale. By comparing our measurements to classical and quantum mechanical calculations with or without inclusion of the Coulomb force, we could identify the deviations of the measurements from predictions of the SFA. In particular, we investigated the different influence of the Coulomb potential on wavepackets that leave the ion core directly and on those that are driven back to ion by the laser field and recollide with it.

6.4 Laser Subcycle Interference Structures

Attosecond electronic restructuring dynamics in atoms and molecules takes place via bound wavepackets formed by coherent superposition of two or more electronic states. A complete characterization of this dynamics necessitates the measurement of the evolution of the relative phases of all states that form the wavepacket. Using the soft X-ray radiation emitted by the process of electron recollision [3] in strong laser fields, several methods capable of gaining access to the relative phases have been developed [12–15, 49]. Here we show an alternative approach based on interference of pairs of electronic wavepackets [16–18].

We have described in the introduction that interference of electron wavepackets released by strong-field ionization during different cycles of period T of an intense laser field leads to ATI peaks spaced by the laser photon energy in photoelectron energy spectra [23]. These types of interferences have also been called the intercycle structures [24]. Wavepackets created within the same laser cycle lead to subcycle interference fringes that are visible as a modulation of the ATI-peak structures. Their relative subcycle release time Δt is encoded in momentum with smaller Δt interfering at higher momenta. See Fig. 6.8 for a schematic.

Interference structures created by two coherently combined optical or matter waves are determined by their relative phases. When pairs of electronic wavepackets are launched at different instants of time, differences in quantum phases accumulated during ionization and propagated in the presence of both the ionic and laser field can be observed as interference patterns in energy or momentum spectra. It is thus possible to actively control and modify the interference fringes with cycle-sculpted laser fields. Using the above described sculpted two-color laser fields, we can exploit this opportunity: Varying the relative phase φ of the two colors allows to sculpt the ionizing field and hence to control the emission times and motion of the interfering wavepackets on an attosecond timescale.

6.4.1 Controlling Subcycle Interference Structures

Interference structures of intercycle (ATI peaks) and subcycle origin are contained in the measured momentum spectra along p_z as a function of φ shown in Fig. 6.3. They are visible in the spectral line-outs in Fig. 6.5a and c for $\varphi = \pi/2$ and $\varphi = 0$. For

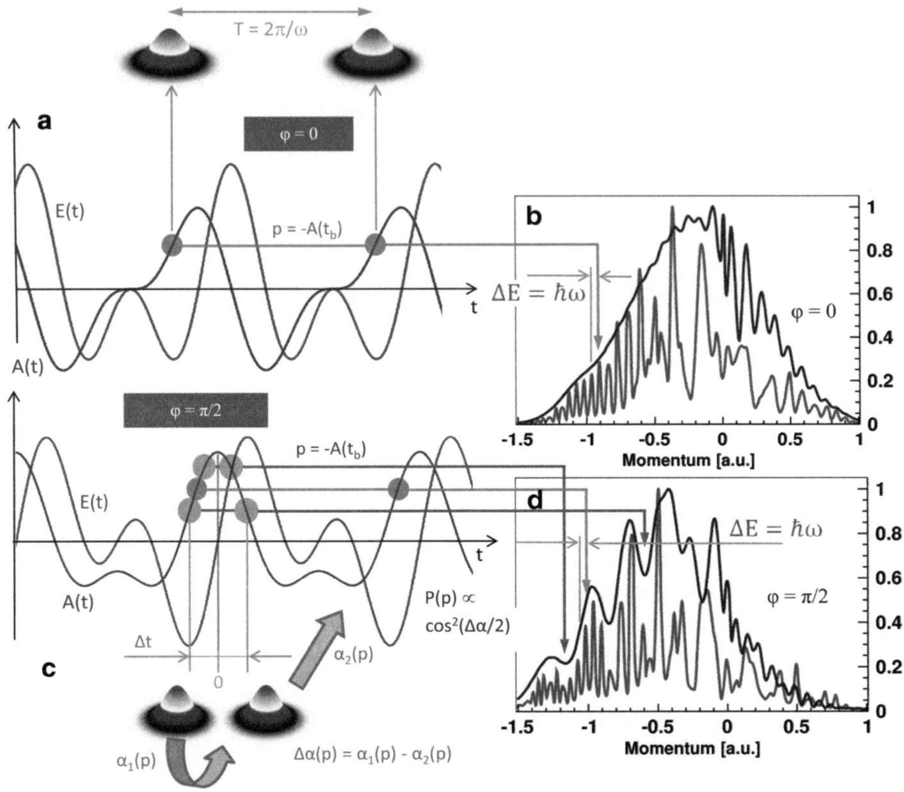

Fig. 6.8 (**a**, **c**) A free electron born at some time t_b within the pulse reaches a final momentum given by the negative vector potential at birth time, $p = -A(t_b)$. Electron wavepackets that reach the same final momentum interfere in momentum space. The same momentum can be reached by wavepackets created during different cycles or by wavepackets created within one cycle. The former lead to interference fringes strictly separated by the photon energy $\hbar\omega = 2\pi/T$, with T the laser-cycle duration, and are usually called ATI peaks [23] [see (**b**) and (**d**)]. The latter lead to interference fringes of subcycle origin and are visible as a modulation of the ATI-peak structures [see (**d**)]. The wavepackets' relative release time Δt is encoded in momentum with smaller Δt interfering at higher momenta. (**b**, **d**) Electron momentum spectra calculated by solving the TDSE for a multicycle (*light lines*) and a single-cycle (*dark lines*) pulse. With a single-cycle pulse, the field periodicity necessary for ATI peaks is broken, and only the subcycle interferences remain visible. A two-color pulse permits to create strongly asymmetric spectra [cf. shape of (**d**)] and therewith to observe the subcycle interference fringes at higher momenta than with carrier-envelope phase controlled few-cycle pulses [7, 25, 26] and free from low-energy resonances [50]. Also, by varying φ, the timing of the wavepackets' creation as well as their motion after ionization can be controlled. Since in interferometry tiny changes to the phase of one of the pulses have a dramatic effect on the observed interferograms, variation of φ thus enables us to clearly distinguish the subcycle interference fringes from ATI peaks. For $\varphi = 0$, the spectrum is expected to be centered at $p = 0$. The observed asymmetry in (**b**) is due to additional momentum that is imparted to the escaping electron by the ionic Coulomb potential

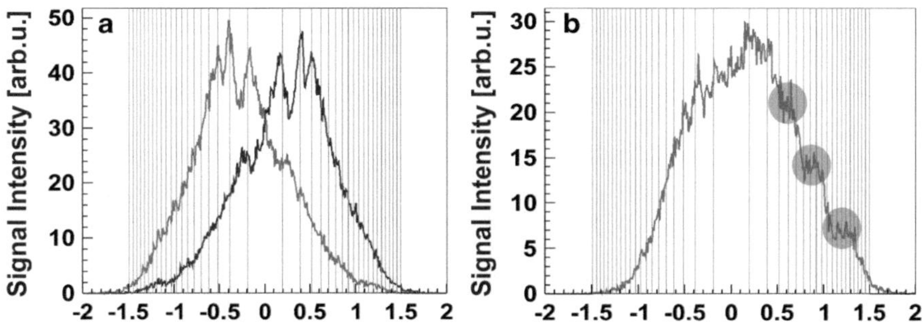

Fig. 6.9 (**a**) Measured momentum spectra of He^+ along the laser polarization direction, z, for $|p_y| < 0.2$ au with p_x integrated over for $\varphi = 0$ (*dark line*) and $\varphi = \pi$ (*light line*). Due to momentum conservation, the ion momentum distribution is the mirrored electron momentum distribution. The vertical lines mark the position of the ATI peak of order n, $p_{n,\mathrm{ATI}} = \pm\sqrt{2(n\omega - I_\mathrm{p} - U_\mathrm{p})}$, where I_p is the ionization potential and U_p the ponderomotive potential. (**b**) Same as (**a**) but for $\varphi = \pi/2$. The dots mark the subcycle interference patterns

$\varphi = \pi/2$, a steplike modulation at $p_z > 0$ is apparent; for $\varphi = 0$, a series of weakly visible peaks can be seen at $p_z < 0$. The visibility of these interference signatures can be enhanced by restricting the electron wavepacket motion to directions along the laser polarization direction z. This can be done by confining the momentum component orthogonal to p_z to small values $|p_y| < 0.2$. The third dimension p_x has been integrated over. We show the resulting spectra for $\varphi = 0/\pi$ and $\varphi = \pi/2$ in Fig. 6.9a and b, respectively. The spectra display interference peaks of different origin and whose position and shape strongly depend on φ. While at small $|p_z|$, i.e., small energies, they coincide with the position of interference peaks equispaced in $\hbar\omega$, commonly known as ATI peaks [23, 24], for larger momenta $|p_z| \gtrsim 1$ au, the experimental resolution is insufficient to resolve them. Instead, the spectrum for $\varphi = \pi/2$ features a modulation of the spectrum unrelated to ATI. These terrace-like structures are the subcycle interference fringes which can be described as an interferogram of the form $P(p_z) \propto \cos^2(\Delta\alpha(p_z)/2)$ created by the phase-difference $\Delta\alpha(p_z)$ of two electron wavepackets released within a single laser cycle; see Fig. 6.8.

The position and shape of the subcycle interference peaks sensitively depend on the shape of the laser field cycle which we control by varying φ. By contrast, the ATI peaks are created by interference of wavepackets released during different laser cycles, and their positions thus reflect the periodicity T of the field giving equispaced peaks in $\hbar\omega$ independent of φ. Subcycle and ATI fringes can be clearly separated from each other by studying the variation of the longitudinal spectrum with φ; see Fig. 6.10a. The subcycle fringes appear as bow-like structures whose positions vary strongly with φ. The strong asymmetry of the spectra about $p_z = 0$, which is a further consequence of the two-color field, allows to detect them well apart from low-energy resonances [50]. Maximum fringe spacing and highest momenta are reached for $\varphi = (0.5 + n)\pi$, $n \in \mathbb{Z}$. In contrast, the position of the ATI

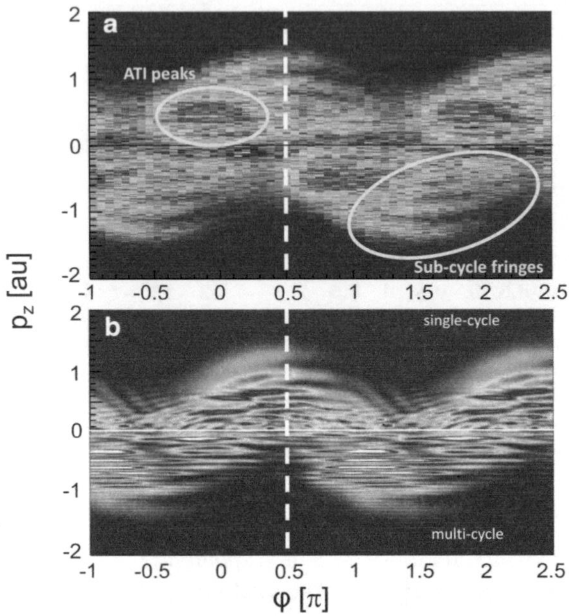

Fig. 6.10 (**a**) Measured momentum spectra of He^+ along p_z, for $|p_y| < 0.2$ au with p_x integrated over as a function of φ. From the raw spectra [as shown in Fig. 6.9a and b], an overall Gaussian shape has been subtracted to enhance the visibility of the subcycle interference patterns that sensitively depend on the laser field shape and therewith on φ, visible as the indicated bow-like structures. The ATI peaks do not depend on the field's shape and are therewith independent of φ. (**b**) Same as (**a**) but calculated using the TDSE (no subtraction was performed). The upper hemisphere shows spectra for a single-cycle pulse, the lower hemisphere for a multicycle pulse

peaks is independent of φ only determined by T (or, equivalently, ω). To enhance the visibility of the subcycle interference structures in Fig. 6.10a, a Gaussian fit function to the spectra multiplied by 0.6 has been subtracted from them for each value of φ. This subtraction enhances the visibility of the steps in the raw spectrum that correspond to the subcycle interference fringes.

6.4.2 Retrieval of Wavepacket Dynamics from Interference Structures

We compare the measured interference structures to simulated ones, shown in Fig. 6.10b. We calculated electron momentum spectra by solving the TDSE in three spatial dimensions in cylindrical coordinates in the velocity gauge within the single-active electron approximation for a model potential [43] of helium. We used pulses of multicycle and single-cycle duration. The simulated spectra are shown in the lower and upper part of Fig. 6.10b, respectively. For the multicycle pulses,

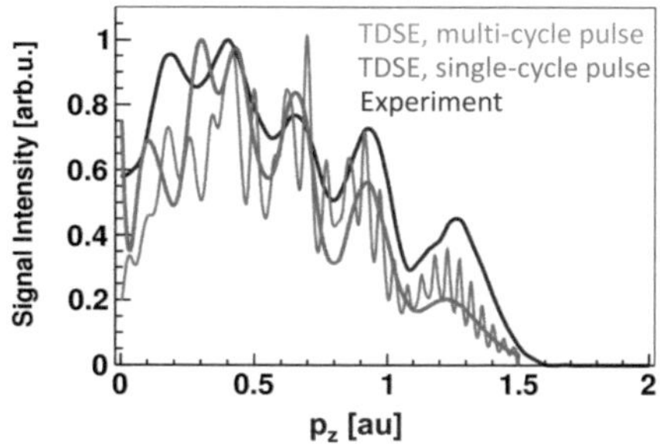

Fig. 6.11 Comparison of normalized spectra for $\varphi = \pi/2$, calculated using the TDSE for a multicycle two-color pulse (*thin light line*) and a single-cycle two-color pulse (*thick medium light line*). The dark line denotes the normalized measured spectrum

they display both the closely spaced ATI peaks and the more widely spaced bow-like subcycle fringes. By reducing the duration of the pulse to a single cycle T, the intercycle interference or ATI peaks can be switched off, while the subcycle interference remains intact.

A detailed comparison between experiment and simulation using the TDSE can be found in Fig. 6.11 for $\varphi = \pi/2$, for which the subcycle fringes reach the highest momenta and largest fringe spacing. In the figure, the experimental spectrum (dark line) is compared to simulated spectra for a multicycle (thin light line) and single-cycle (thick medium light line) pulse. The spectrum for the multicycle pulse consists of a series of ATI peaks equidistant in energy, i.e., with decreasing spacing towards higher momenta. The ATI peak structure is clearly modulated with minima and maxima following those of the spectrum for a single-cycle pulse. To facilitate a quantitative comparison of simulation with experiment, the calculated spectra were averaged over an intensity range from $0.5I_0$ to I_0, with I_0 being the peak intensity of the laser pulse, to account for intensity smearing in the laser focal volume. The averaging leads to a reduced fringe modulation depth and a slight momentum shift of the fringe peaks by $\Delta p_z \approx -0.04$ au. Comparison of the experimental spectrum with that created by the single-cycle pulse clearly confirms the appearance of the subcycle peaks in the experiment while the more densely spaced ATI peaks are not resolved.

From the measured interferogram at fixed φ, $P(p_z, \varphi) \propto \cos^2(\Delta\alpha(p_z)/2)$, it is possible to extract the phase-difference $\Delta\alpha(p_z)$ accumulated by two wavepackets that are released within one optical cycle of the two-color pulse. By using the semiclassical connection [29] between the momentum p_z and the time of birth of the wavepacket, the phase difference $\Delta\alpha(p_z)$ can be expressed as a function of the time delay between the emission of the two wavepackets within the same cycle; see Fig. 6.8. Using the relation $\mathbf{p} = -\mathbf{A}(t_b)$, with $\mathbf{A}(t)$ the laser vector potential, that we used above in Sect. 6.3.1 allows t_b to be determined by inversion. However, this approximation neglects the force of the Coulomb binding potential on released wavepackets and only considers the force of the laser electric field.

The mapping of p_z onto Δt strongly depends on φ. Figure 6.12 visualizes the dependence of momentum to subcycle time-mapping on the shape of the sculpted two-color pulse. A maximum range of momenta is covered by a pulse with φ around $\pi/2$ giving rise to subcycle time slits 50 as $\lesssim \Delta t \lesssim$ 1 fs. Note that time

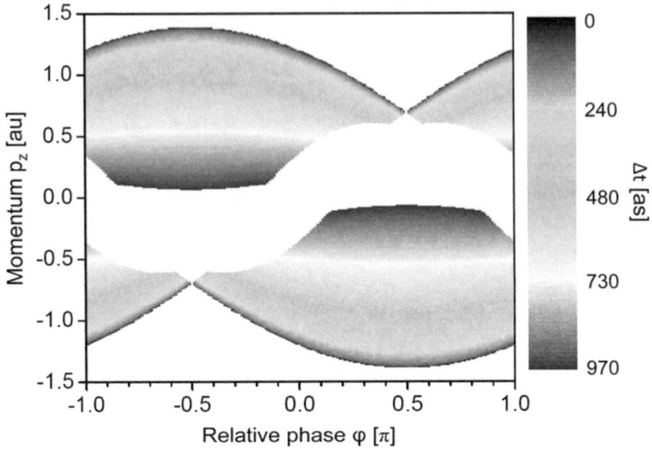

Fig. 6.12 Dependence of momentum to subcycle time-mapping on the shape of a sculpted two-color pulse. Difference Δt in attoseconds between the release times of two wavepackets (encoded in *gray scale*) as a function of momentum at which they interfere for different relative phases φ of the two-color pulse used in the experiments. The *white area* indicates momenta where either no or only one of the two packets reaches this momentum. The largest range of p_z and Δt appears for a sculpted laser field with $\varphi = \pi/2$

scales below $\approx$50 as are effectively inaccessible since ionization is suppressed at small fields. Moreover, the useful time window is limited to reasonably high momenta $\gtrsim$0.3 au or $\Delta t \lesssim 0.8$ fs due to the failure of the SFA for small momenta, for which the influence of the Coulomb potential and low-energy resonances becomes increasingly important. For the present two-color sculpted pulse, delay times between 50 as $\lesssim \Delta t \lesssim$ 800 as can be probed with an estimated precision of less than 10 as at the largest delay times.

The such obtained phase-difference $\Delta\alpha(\Delta t)$ contains contributions due to the interaction with the laser field and with the Coulomb potential. The contribution of the laser field is the well-known Volkov phase and that due to the Coulomb field is the Coulomb eikonal phase [30]. If both of them are subtracted from the measured phase $\Delta\alpha(\Delta t)$, we are left with the subcycle phase evolution of the bound state $\Delta\alpha_B(\Delta t)$. Thus, our work shows that measurement of subcycle electron wavepacket interference patterns can serve as a tool to measure the dynamics of the valence electron cloud in atoms and molecules, complementary to high-harmonic spectroscopy [12, 13, 15, 51] and single attosecond pulse spectroscopy [52, 53].

6.5 Conclusion

We generated cycle-sculpted two-color waveforms to experimentally drive electronic wavepackets emitted by strong-field ionization from helium, neon, and argon gas atoms and analyzed their momentum spectra measured by the COLTRIMS technique. Varying the relative phase φ of the two colors allowed us to sculpt the ionizing field and hence to control the emission times and motion of the interfering wavepackets on an attosecond timescale. Using semiclassical calculations, we investigated the influence of the ionic Coulomb field onto the motion of the wavepackets and identified regimes where the widely used strong-field approximation leads to severe errors in the interpretation of typical probe signals such as electron momenta or emission times of photons. We further showed that the measured electron momentum spectra contain interference patterns created by pairs of electron wavepackets that are released with delays smaller than the laser oscillation period. Our cycle-sculpted laser field allowed us to experimentally distinguish these subcycle structures from interference fringes generated by wavepackets that are released during different cycles, which are observed as ATI peaks in photoelectron energy spectra spaced by the laser photon energy $\hbar\omega$. We argue that the measured subcycle interferograms can be used to extract the subcycle phase evolution of the laser-driven complex bound-state wavefunction.

Acknowledgements We acknowledge, funding by the Austrian Science Fund (FWF) under grants P21463-N22 and SFB-016 and discussions with J. Burgdörfer, D. Arbó, E. Persson, and S. Gräfe.

References

1. P.B. Corkum, F. Krausz, Nat. Phys. **3**(6), 381 (2007). DOI 10.1038/nphys620. URL http://www.nature.com/doifinder/10.1038/nphys620
2. F. Krausz, M. Ivanov, Rev. Mod. Phys. **81**(1), 163 (2009). DOI 10.1103/RevModPhys.81.163. URL http://link.aps.org/doi/10.1103/RevModPhys.81.163
3. P. Corkum, Phys. Rev. Lett. **71**(13), 1994 (1993). DOI 10.1103/PhysRevLett.71.1994. URL http://link.aps.org/doi/10.1103/PhysRevLett.71.1994
4. M. Kitzler, M. Lezius, Phys. Rev. Lett. **95**(25), 253001 (2005). DOI 10.1103/PhysRevLett.95.253001. URL http://link.aps.org/doi/10.1103/PhysRevLett.95.253001
5. M. Kitzler, X. Xie, A. Scrinzi, A. Baltuska, Phys. Rev. A **76**(1), 011801 (2007). DOI 10.1103/PhysRevA.76.011801. URL http://link.aps.org/doi/10.1103/PhysRevA.76.011801
6. M. Spanner, O. Smirnova, P.B. Corkum, M.Y. Ivanov, J. Phys. B Atom. Mol. Opt. Phys. **37**(12), L243 (2004). DOI 10.1088/0953-4075/37/12/L02. URL http://stacks.iop.org/0953-4075/37/i=12/a=L02?key=crossref.d453c23ea94b6703d9e16eeda244ce36
7. F. Lindner, M. Schätzel, H. Walther, A. Baltuška, E. Goulielmakis, F. Krausz, D. Milošević, D. Bauer, W. Becker, G. Paulus, Phys. Rev. Lett. **95**(4), 040401 (2005). DOI 10.1103/PhysRevLett.95.040401. URL http://link.aps.org/doi/10.1103/PhysRevLett.95.040401
8. S. Yurchenko, S. Patchkovskii, I. Litvinyuk, P. Corkum, G. Yudin, Phys. Rev. Lett. **93**(22), 223003 (2004). DOI 10.1103/PhysRevLett.93.223003. URL http://link.aps.org/doi/10.1103/PhysRevLett.93.223003
9. M. Meckel, D. Comtois, D. Zeidler, A. Staudte, D. Pavicic, H.C. Bandulet, H. Pépin, J.C. Kieffer, R. Dörner, D.M. Villeneuve, P.B. Corkum, Science (New York, N.Y.) **320**(5882), 1478 (2008). DOI 10.1126/science.1157980. URL http://www.ncbi.nlm.nih.gov/pubmed/18556555
10. Y. Huismans, A. Rouzée, A. Gijsbertsen, J.H. Jungmann, A.S. Smolkowska, P.S.W.M. Logman, F. Lépine, C. Cauchy, S. Zamith, T. Marchenko, J.M. Bakker, G. Berden, B. Redlich, A.F.G. van der Meer, H.G. Muller, W. Vermin, K.J. Schafer, M. Spanner, M.Y. Ivanov, O. Smirnova, D. Bauer, S.V. Popruzhenko, M.J.J. Vrakking, Science (New York, N.Y.) **331**, 61 (2010). DOI 10.1126/science.1198450. URL http://www.ncbi.nlm.nih.gov/pubmed/21163963

11. S. Baker, J.S. Robinson, C.A. Haworth, H. Teng, R.A. Smith, C.C. Chirila, M. Lein, J.W.G. Tisch, J.P. Marangos, Science **312**(5772), 424 (2006). DOI 10.1126/science.1123904. URL http://www.ncbi.nlm.nih.gov/pubmed/16513942
12. O. Smirnova, Y. Mairesse, S. Patchkovskii, N. Dudovich, D. Villeneuve, P. Corkum, M.Y. Ivanov, Nature **460**(7258), 972 (2009). DOI 10.1038/nature08253. URL http://www.ncbi.nlm.nih.gov/pubmed/19626004
13. S. Haessler, J. Caillat, W. Boutu, C. Giovanetti-Teixeira, T. Ruchon, T. Auguste, Z. Diveki, P. Breger, A. Maquet, B. Carré, R. Taïeb, P. Salières, Nat. Phys. **6**(3), 200 (2010). DOI 10.1038/nphys1511. URL http://www.nature.com/doifinder/10.1038/nphys1511
14. H.J. Wörner, J.B. Bertrand, D.V. Kartashov, P.B. Corkum, D.M. Villeneuve, Nature **466**(7306), 604 (2010). DOI 10.1038/nature09185. URL http://www.nature.com/doifinder/10.1038/nature09185
15. Y. Mairesse, J. Higuet, N. Dudovich, D. Shafir, B. Fabre, E. Mével, E. Constant, S. Patchkovskii, Z. Walters, M.Y. Ivanov, O. Smirnova, Phys. Rev. Lett. **104**(21), 213601 (2010). DOI 10.1103/PhysRevLett.104.213601. URL http://link.aps.org/doi/10.1103/PhysRevLett.104.213601
16. M. Wollenhaupt, A. Assion, D. Liese, C. Sarpe-Tudoran, T. Baumert, S. Zamith, M. Bouchene, B. Girard, A. Flettner, U. Weichmann, G. Gerber, Phys. Rev. Lett. **89**(17), 173001 (2002). DOI 10.1103/PhysRevLett.89.173001. URL http://link.aps.org/doi/10.1103/PhysRevLett.89.173001
17. T. Remetter, P. Johnsson, J. Mauritsson, K. Varjú, Y. Ni, F. Lépine, E. Gustafsson, M. Kling, J. Khan, R. López-Martens, K.J. Schafer, M.J.J. Vrakking, A. L'Huillier, Nat. Phys. **2**(5), 323 (2006). DOI 10.1038/nphys290. URL http://www.nature.com/doifinder/10.1038/nphys290
18. F. Quéré, J. Itatani, G. Yudin, P. Corkum, Phys. Rev. Lett. **90**(7), 073902 (2003). DOI 10.1103/PhysRevLett.90.073902. URL http://link.aps.org/doi/10.1103/PhysRevLett.90.073902
19. T. Weinacht, J. Ahn, P. Bucksbaum, Phys. Rev. Lett. **80**(25), 5508 (1998). DOI 10.1103/PhysRevLett.80.5508. URL http://link.aps.org/doi/10.1103/PhysRevLett.80.5508
20. J. Mauritsson, T. Remetter, M. Swoboda, K. Klünder, A. L'Huillier, K. Schafer, O. Ghafur, F. Kelkensberg, W. Siu, P. Johnsson, M. Vrakking, I. Znakovskaya, T. Uphues, S. Zherebtsov, M. Kling, F. Lépine, E. Benedetti, F. Ferrari, G. Sansone, M. Nisoli, Phys. Rev. Lett. **105**(5), 053001 (2010). DOI 10.1103/PhysRevLett.105.053001. URL http://link.aps.org/doi/10.1103/PhysRevLett.105.053001
21. D. Fittinghoff, J. Bowie, J. Sweetser, R. Jennings, M. Krumbüugel, K. DeLong, R. Trebino, I. Walmsley, Opt. Lett. **21**(12), 884 (1996). URL http://www.opticsinfobase.org/abstract.cfm?URI=ol-21-12-884
22. C. Iaconis, V. Wong, I. Walmsley, IEEE J. Sel. Top. Quant. Electron. **4**(2), 285 (1998). DOI 10.1109/2944.686734. URL http://ieeexplore.ieee.org/lpdocs/epic03/wrapper.htm?arnumber=686734
23. P. Agostini, F. Fabre, G. Mainfray, G. Petite, N. Rahman, Phys. Rev. Lett. **42**(17), 1127 (1979). DOI 10.1103/PhysRevLett.42.1127. URL http://link.aps.org/doi/10.1103/PhysRevLett.42.1127
24. D.G. Arbó, K.L. Ishikawa, K. Schiessl, E. Persson, J. Burgdörfer, Phys. Rev. A **81**(2), 021403 (2010). DOI 10.1103/PhysRevA.81.021403. URL http://link.aps.org/doi/10.1103/PhysRevA.81.021403
25. D. Arbó, E. Persson, J. Burgdörfer, Phys. Rev. A **74**(6), 063407 (2006). DOI 10.1103/PhysRevA.74.063407. URL http://link.aps.org/doi/10.1103/PhysRevA.74.063407
26. R. Gopal, K. Simeonidis, R. Moshammer, T. Ergler, M. Dürr, M. Kurka, K.U. Kühnel, S. Tschuch, C.D. Schröter, D. Bauer, J. Ullrich, A. Rudenko, O. Herrwerth, T. Uphues, M. Schultze, E. Goulielmakis, M. Uiberacker, M. Lezius, M. Kling, Phys. Rev. Lett. **103**(5), 053001 (2009). DOI 10.1103/PhysRevLett.103.053001. URL http://link.aps.org/doi/10.1103/PhysRevLett.103.053001
27. D. Schumacher, F. Weihe, H. Muller, P. Bucksbaum, Phys. Rev. Lett. **73**(10), 13441347 (1994). DOI doi/10.1103/PhysRevLett.73.1344. URL http://link.aps.org/doi/10.1103/PhysRevLett.73.1344

28. H. Muller, P. Bucksbaum, D. Schumacher, A. Zavriyev, J. Phys. B Atom. Mol. Opt. Phys. **23**, 2761 (1990). DOI 10.1088/0953-4075/23/16/018. URL http://iopscience.iop.org/0953-4075/23/16/018
29. M. Lewenstein, K. Kulander, K. Schafer, P. Bucksbaum, Phys. Rev. A **51**(2), 1495 (1995). DOI 10.1103/PhysRevA.51.1495. URL http://link.aps.org/doi/10.1103/PhysRevA.51.1495
30. D. Arbó, K. Ishikawa, K. Schiessl, E. Persson, J. Burgdörfer, Phys. Rev. A **82**(4), 043426 (2010). DOI 10.1103/PhysRevA.82.043426. URL http://link.aps.org/doi/10.1103/PhysRevA.82.043426
31. D. Arbó, S. Yoshida, E. Persson, K. Dimitriou, J. Burgdörfer, Phys. Rev. Lett. **96**(14), 143003 (2006). DOI 10.1103/PhysRevLett.96.143003. URL http://link.aps.org/doi/10.1103/PhysRevLett.96.143003
32. S. Chelkowski, A. Bandrauk, A. Apolonski, Phys. Rev. A **70**(1), 013815 (2004). DOI 10.1103/PhysRevA.70.013815. URL http://link.aps.org/doi/10.1103/PhysRevA.70.013815
33. S. Chelkowski, A. Bandrauk, Phys. Rev. A **71**(5), 053815 (2005). DOI 10.1103/PhysRevA.71.053815. URL http://link.aps.org/doi/10.1103/PhysRevA.71.053815
34. O. Smirnova, M. Spanner, M. Ivanov, Phys. Rev. A **77**(3), 033407 (2008). DOI 10.1103/PhysRevA.77.033407. URL http://link.aps.org/doi/10.1103/PhysRevA.77.033407
35. F. Ehlotzky, Phys. Rep. **345**(4), 175 (2001). DOI 10.1016/S0370-1573(00)00100-9. URL http://linkinghub.elsevier.com/retrieve/pii/S0370157300001009
36. H. Ohmura, T. Nakanaga, M. Tachiya, Phys. Rev. Lett. **92**(11), 113002 (2004). DOI 10.1103/PhysRevLett.92.113002. URL http://link.aps.org/doi/10.1103/PhysRevLett.92.113002
37. H. Ohmura, N. Saito, M. Tachiya, Phys. Rev. Lett. **96**(17), 173001 (2006). DOI 10.1103/PhysRevLett.96.173001. URL http://link.aps.org/doi/10.1103/PhysRevLett.96.173001
38. D. Ray, F. He, S. De, W. Cao, H. Mashiko, P. Ranitovic, K. Singh, I. Znakovskaya, U. Thumm, G. Paulus, M. Kling, I. Litvinyuk, C. Cocke, Phys. Rev. Lett. **103**(22), 223201 (2009). DOI 10.1103/PhysRevLett.103.223201. URL http://link.aps.org/doi/10.1103/PhysRevLett.103.223201
39. S. De, I. Znakovskaya, D. Ray, F. Anis, N. Johnson, I. Bocharova, M. Magrakvelidze, B. Esry, C. Cocke, I. Litvinyuk, M. Kling, Phys. Rev. Lett. **103**(15), 153002 (2009). DOI 10.1103/PhysRevLett.103.153002. URL http://link.aps.org/doi/10.1103/PhysRevLett.103.153002
40. R. Dörner, V. Mergel, O. Jagutzki, L. Spielberger, J. Ullrich, R. Moshammer, H. Schmidt-Böcking, Phys. Rep. **330**(2–3), 95 (2000). DOI 10.1016/S0370-1573(99)00109-X. URL http://linkinghub.elsevier.com/retrieve/pii/S037015739900109X
41. R. Moshammer, B. Feuerstein, W. Schmitt, A. Dorn, C. Schröter, J. Ullrich, H. Rottke, C. Trump, M. Wittmann, G. Korn, Others, Phys. Rev. Lett. **84**(3), 447 (2000). URL http://link.aps.org/doi/10.1103/PhysRevLett.84.447
42. G. Yudin, M. Ivanov, Phys. Rev. A **64**(1), 013409 (2001). DOI 10.1103/PhysRevA.64.013409. URL http://link.aps.org/doi/10.1103/PhysRevA.64.013409
43. X.M. Tong, C.D. Lin, J. Phys. B Atom. Mol. Opt. Phys. **38**(15), 2593 (2005). DOI 10.1088/0953-4075/38/15/001. URL http://stacks.iop.org/0953-4075/38/i=15/a=001?key=crossref.7d04da6e26e52de9b427ea49c7ca2a83
44. N. Delone, V. Krainov, J. Opt. Soc. Am. B **8**(6), 1207 (1991). URL http://www.opticsinfobase.org/ol/ViewMedia.cfm?id=6026&seq=0
45. P. Ho, J. Eberly, Phys. Rev. Lett. **95**(19), 193002 (2005). DOI 10.1103/PhysRevLett.95.193002. URL http://link.aps.org/doi/10.1103/PhysRevLett.95.193002
46. P. Ho, R. Panfili, S. Haan, J. Eberly, Phys. Rev. Lett. **94**(9), 093002 (2005). DOI 10.1103/PhysRevLett.94.093002. URL http://link.aps.org/doi/10.1103/PhysRevLett.94.093002
47. X. Xie, A. Scrinzi, M. Wickenhauser, A. Baltuška, I. Barth, M. Kitzler, Phys. Rev. Lett. **101**(3), 033901 (2008). DOI 10.1103/PhysRevLett.101.033901. URL http://link.aps.org/doi/10.1103/PhysRevLett.101.033901
48. P. Dietrich, N. Burnett, M. Ivanov, P. Corkum, Phys. Rev. A **50**(5), R3585 (1994). DOI 10.1103/PhysRevA.50.R3585. URL http://link.aps.org/doi/10.1103/PhysRevA.50.R3585
49. X. Zhou, R. Lock, W. Li, N. Wagner, M. Murnane, H. Kapteyn, Phys. Rev. Lett. **100**(7), 073902 (2008). DOI 10.1103/PhysRevLett.100.073902. URL http://link.aps.org/doi/10.1103/PhysRevLett.100.073902

50. R. Freeman, P. Bucksbaum, H. Milchberg, S. Darack, D. Schumacher, M. Geusic, Phys. Rev. Lett. **59**(10), 1092 (1987). DOI 10.1103/PhysRevLett.59.1092. URL http://link.aps.org/doi/10.1103/PhysRevLett.59.1092
51. W. Li, X. Zhou, R. Lock, S. Patchkovskii, A. Stolow, H.C. Kapteyn, M.M. Murnane, Science **322**(5905), 1207 (2008). DOI 10.1126/science.1163077. URL http://www.ncbi.nlm.nih.gov/pubmed/18974317
52. E. Goulielmakis, Z.H. Loh, A. Wirth, R. Santra, N. Rohringer, V.S. Yakovlev, S. Zherebtsov, T. Pfeifer, A.M. Azzeer, M.F. Kling, S.R. Leone, F. Krausz, Nature **466**(7307), 739 (2010). DOI 10.1038/nature09212. URL http://www.nature.com/doifinder/10.1038/nature09212
53. M. Uiberacker, T. Uphues, M. Schultze, A.J. Verhoef, V. Yakovlev, M.F. Kling, J. Rauschenberger, N.M. Kabachnik, H. Schröder, M. Lezius, K.L. Kompa, H.G. Muller, M.J.J. Vrakking, S. Hendel, U. Kleineberg, U. Heinzmann, M. Drescher, F. Krausz, Nature **446**(7136), 627 (2007). DOI 10.1038/nature05648. URL http://www.ncbi.nlm.nih.gov/pubmed/17410167

Chapter 7
Characterization of Femtosecond Laser Filament-Induced Plasma and Its Application to Atmospheric Sensing

HuaiLiang Xu, Ya Cheng, ZhiZhan Xu, and See Leang Chin

Abstract In this chapter, we demonstrate experimental characterizations of plasma of different phase targets induced by femtosecond laser filamentation using optical emission spectroscopy and present an interesting phenomenon of lasing actions occurring in the plasma column in air due to two different mechanisms, i.e., amplified spontaneous emission and self-generated harmonic seeding lasing. Several examples in terms of sensing of atmospheric constituents of interest based on characteristic filament-induced plasma are then given.

7.1 Introduction

The ability to measure atmospheric species over a long distance is of essential importance in dealing with a variety of environmental issues from global warming and stratospheric ozone depletion to early warning of biological and nuclear power plant's leak that threatens public and defense security. To date, because of the capacity of nonintrusiveness, online and real-time sample analysis, various laser-based spectroscopic techniques such as Fourier transform infrared spectroscopy,

H. Xu (✉)
State Key Laboratory on Integrated Optoelectronics, College of Electronic Science and Engineering, Jilin University, Changchun 130012, China
e-mail: huailiang@jlu.edu.cn

Y. Cheng · Z. Xu
State Key Laboratory of High Field Laser Physics, Shanghai Institute of Optics and Fine Mechanics, Chinese Academy of Sciences, Shanghai 201800, China
e-mail: ya.cheng@siom.ac.cn; zzxu@mail.shcnc.ac.cn

S.L. Chin
Center for Optics, Photonics and Laser (COPL) & Department of Physics, Engineering Physics and Optics, Université Laval, Quebec City, Qc G1V 0A6, Canada
e-mail: slchin@phy.ulaval.ca

K. Yamanouchi et al. (eds.), *Progress in Ultrafast Intense Laser Science VIII*, Springer Series in Chemical Physics 103, DOI 10.1007/978-3-642-28726-8_7,
© Springer-Verlag Berlin Heidelberg 2012

differential absorption light detection and ranging (DIAL), and laser-induced fluorescence spectroscopy have attracted great interest; they have demonstrated their versatility in monitoring remote targets including gases, aerosols, and solids with the detection sensitivity in the ppm-ppb levels, depending on the detection distance and the sensed species [1]. Nevertheless, most of the laser remote sensing techniques generally require laser sources to be tuned to a special wavelength for one species at a time. For example, in the case of DIAL, one wavelength is set to an on-resonance absorption peak of the target species under study, while a second wavelength is set to off-resonance of the species. The difference in the backward signals gives the information on the concentration of the species along the optical path. Therefore, sensing more than one species of interest in the atmosphere basically requires multiple laser sources to cover wide spectral range [2].

The ideal situation in the field of laser remote sensing would be that one coherent light source could do all the tasks with the ability to monitor a very broad range of constituents in the atmosphere. The possibility to perform multielemental material analysis with only one laser source could be achieved using the techniques such as laser-induced breakdown spectroscopy (LIBS) and laser photofragmentation spectroscopy (PFS), which are based on the spectral detection of optical emissions from laser-induced plasma. When a pulsed laser with sufficient energy is tightly focused and interacts with a sample, a plasma volume is produced that gives characteristic optical emissions. The properties of the plasma and the material compounds can be determined from the measured emission spectra. The key aspect of such techniques applied to the field of remote sensing is to ensure that sufficiently high laser *intensities* are delivered to a remote distance to induce a plasma breakdown or photofragmentation of the sample resulting in characteristic optical emissions. Clearly, because of diffraction, it is difficult to project the nanosecond laser pulses with high intensities onto a remote sample, leading to limited operation ranges of the LIBS or PF techniques using nanosecond laser systems.

Advances in high-power femtosecond laser technologies make it possible to use only one laser source for simultaneously sensing multipollutants in the kilometer range [3, 4]. When femtosecond laser pulses with the power beyond a threshold value (critical power) propagate in air, due to the optical Kerr effect the laser beam self-focuses, leading to multiphoton/tunnel ionization of air molecules. This produces a plasma that defocuses the laser beam. The dynamic balance between Kerr self-focusing and defocusing in air limits the laser intensity to about $5 \times 10^{13}\,\mathrm{W/cm^2}$ (intensity clamping) in the focal spot when the 800-nm laser pulses are employed and leaves a long weak plasma column in the laser propagation path, a so-called filament [2,5–7]. Femtosecond filamentation has been shown to reach remote distances as far as a few kilometers. Filaments can propagate through an adverse atmospheric condition such as fog and clouds better than normal nonfilamenting laser beam due to the energy reservoir surrounding the filament cores [8–11]. A very-low-energy fluctuation and an excellent mode quality inside the filament core can be achieved due to intensity clamping and self-filtering in the filamentation process [12]. In particular, the high clamped intensity inside the filament core can induce all matters in the path of filament fragmentation, resulting in characteristic optical emissions from the excited fragments. Such properties of the filaments make them well suited for atmospheric sensing [13].

The filament-induced plasmas have some very unique properties. For example, they are much longer than the Rayleigh length and could be extended up to hundreds of meters. The high clamped intensity inside the filament core leads to a stable signal at different locations of the interaction of filament with materials. So far, many methods such as optical interferometry [14], plasma conductivity [15, 16], and longitudinal diffraction [17] have been utilized to characterize the physical parameters of the plasma induced by femtosecond laser filamentation, but here, we will only focus on the characterization of the filament-induced plasma by optical emission spectroscopy performed in our previous studies.

In this chapter, we will first briefly introduce the principle of the optical emission spectroscopy for characterizing the plasma. It is followed by our representative results on the characterization of filament-induced plasma of solid targets as well as that in air at atmospheric pressure. We will then give our attention to remote lasing actions occurring in the plasma column. Finally, several examples in terms of atmospheric sensing based on filament-induced nonlinear spectroscopy are overviewed

7.2 Optical Emission Spectroscopy for Characterizing Filamentation-Induced Plasma

The optical emission spectroscopy of laser induced-plasma is a convenient and reliable measurement technique [18]. Although the physical properties of laser-induced plasma are strongly dependent on experimental conditions, the physics behind the characterization of laser-induced plasma by optical emission spectroscopy is universal. That is, laser-induced plasma is treated as a spectroscopic source that emits different spectral lines/bands and continuum radiation. By analyzing these optical emissions, one can obtain characteristic physical parameters such as the electron density and temperature and consequently characterizing the laser-induced plasma.

The principles of the optical emission spectroscopy for characterizing laser-induced plasma have been reviewed extensively, e.g., in [18, 19]. In this method, the plasma should hold the condition of local thermodynamic equilibrium (LTE), which can be judged by the McWhirter criterion [20]

$$N_{\mathrm{e}}\,(\mathrm{cm}^{-3}) \geq N_{\mathrm{cr}} = 1.6 \times 10^{12}\sqrt{T}(\Delta E)^3, \tag{7.1}$$

where N_{cr} is the lower limit of the electron density (N_{e}); ΔE, in eV, is the energy difference; and T, in $^\circ K$, the plasma temperature. When the plasma satisfies the LTE condition, an analysis of the line intensity versus the energy of the upper level of the transition can be performed to determine the plasma temperature since the population density of the excited electronic state can be described by the well-known

Boltzmann distribution. The line intensity related to the plasma temperature (T) can be expressed as

$$I_{ki} = \frac{1}{4\pi}\frac{hc}{\lambda}\frac{N_0}{U(T)}g_k A_{ki}\exp\left(-\frac{E_k}{kT}\right), \tag{7.2}$$

where $U(T)$ is the partition function of the species, N_0 the number density of the species, E_k the excitation energy, g_k the statistical weight of the upper level k, I_{ki} and A_{ki} are the intensity and probability for the transition from the upper level k to the lower level i, and λ, h, and c are the transition wavelength, Planck's constant, and light speed, respectively. Logarithmizing (7.2) yields

$$\ln\left(\frac{\lambda I_{ki}}{g_k A_{ki}}\right) \propto -\frac{1}{kT}E_k. \tag{7.3}$$

Thus, from the slope $-1/kT$ of the so-called Boltzmann plot, $\ln(\lambda I_{ki}/g_k A_{ki})$ versus E_k, the plasma temperature T can be obtained.

On the other hand, the analysis of the spectral line width can determine the electron density, which is based on the Stark effect in the plasma. The collisions of the emitting particles with electrons and ions in the plasma results in Stark line broadening with the full width half maximum (FWHM) ($\Delta\lambda_{1/2}$) expressed as

$$\Delta\lambda_{1/2} = 2\omega\left(\frac{N_e}{10^{16}}\right) + 3.5A\left(\frac{N_e}{10^{16}}\right)^{5/4}\left[1 - \frac{3}{4}N_D^{-1/3}\right]\omega, \tag{7.4}$$

where N_e is the electron density, ω is the parameter representing the contribution from electron impact, A is the parameter representing the contribution from the collision with ions, and N_D is the number of particles in the Debye sphere, which can be estimated by $N_D = 1.72 \times 109\, T^{3/2} N_e^{-1/2}$[18]. The first term in (7.4) refers to broadening due to the electron impact, whereas the second term provides the correction originating from the collision with ions.

In addition, the electron density may be determined from the Stark shift of the spectral lines, and the plasma temperature may be obtained from the Saha–Bolzmann equation if the spectral lines selected are from successive ionization stages of the same element. The readers are referred to [18–20] for more discussions about this technique.

7.3 Physical Properties of Filamentation Induced Plasma of Solid Targets

In this section, we will focus on the characterization of plasma produced by the interaction of femtosecond laser filament with a lead solid target [21] and discuss the advantages of filament-induced breakdown spectroscopy (FIBS) for remote sensing.

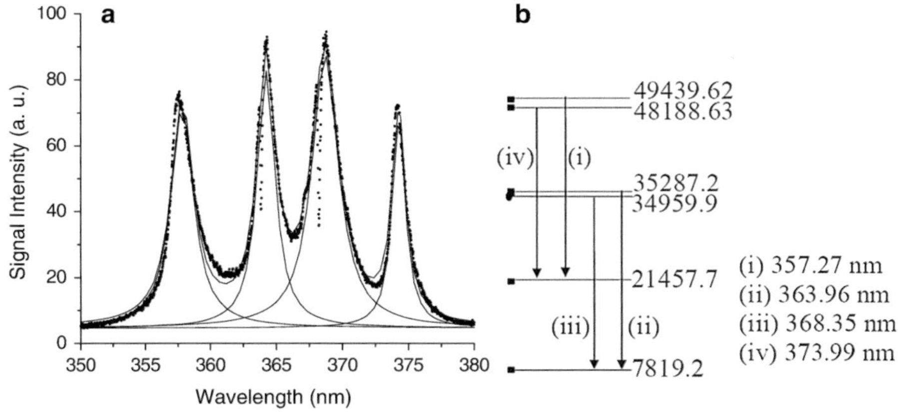

Fig. 7.1 (**a**) Spectrum of the filament-induced lead plasma with a delay time of 40 ns (*dotted*) and multi-peak Lorentzian fitting (*solid lines*) and (**b**) Partial energy diagram of Pb I and corresponding transition wavelengths

Figure 7.1a shows the emission spectrum of filament-induced lead plasma in the region of 350–380 nm. The spectral lines were assigned to atomic Pb transitions (Fig. 7.1b) [22]. In this measurement, the laser pulses were focused using a fused silica lens ($f = 500$ cm) in air to generate a single filament. The laser pulse had an energy of 5 mJ and a pulse duration of about 42 fs. The pulse spectrum was centered at 800 nm. The lead sample was fixed on a rotating stage and placed 4 m away from the lens. This is roughly at the central part of the filament [2]. The FIBS spectrum was obtained by a 0.5-m spectrometer (Acton Research Corp., SpectraPro-500i) equipped with a gated intensified charge-coupled device (ICCD, Princeton instruments Pi-Max 512). The spectrum was recorded with an ICCD gate width of 20 ns and a delay time of 40 ns with respect to the interaction time of filament with the target.

According to (7.3), the plasma temperature was estimated from the emission intensities of three lead (Pb I) lines at 357.27, 363.96, and 373.99 nm. The corresponding transition probabilities of A_{ki} ($10^6\,s^{-1}$) = 95.3, 30.8, and 53.3 and the statistical weights of the upper states of $g_k = 3$, 3, and 5, were adopted, respectively [23]. Figure 7.2a shows, as an example, the Boltzmann plots for a time delay of 40 ns, giving rise to a temperature of 6,421 K according to the slope obtained from the linear fitting. The plasma temperatures for the early stage emission of the plasma, with delay time less than 100 ns, were shown in Fig. 7.2b. To check whether the filament-induced plasma holds the LTE condition, we substituted the highest temperature of 6,794 K obtained for a delay time of 20 ns and the energy difference of $\Delta E = 3.47$ eV for the 7s $^1P_1 \rightarrow 6p^{2\,1}D_2$ transition at 357.27 nm into (7.1); N_{cr} was determined to be the order of $\sim 10^{15}\,cm^{-3}$, which is much smaller than $N_e \sim 10^{17} cm^{-3}$ obtained here (see below), and thus the condition of LTE was justified.

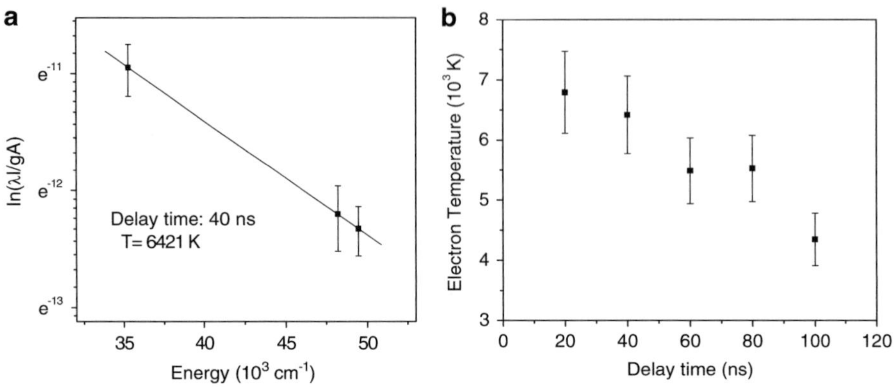

Fig. 7.2 (**a**) Typical Boltzmann plot obtained with a delay time of 40 ns and (**b**) the plasma temperature as a function of the delay time

The electron density of the plasma was estimated from (7.4). For the spectral line at 373.99 nm, the values of ω and A are $\omega = 0.014$ nm and $A = 0.0082$ nm, respectively [24]. With these values, the broadening originating from the collision with ions was estimated to be negligible as the plasma temperature of $T = 6,500$ K, and the electron density of $N_e = 10^{18}\,\text{cm}^{-3}$ was adopted; thus, this contribution could be neglected, and thus (7.4) becomes

$$\Delta\lambda_{1/2} \cong 2\omega\left(\frac{N_e}{10^{16}}\right). \tag{7.5}$$

Using (7.5), the electron density of $N_e \approx 8 \times 10^{17}\,\text{cm}^{-3}$ was obtained for a 20-ns time delay with respect to the laser pulse arriving on the target and a 20-ns gate width. By setting the gate width to 20 ns and changing the time delay, the temporal behavior of the electron density was obtained and shown in Fig. 7.3.

The electron density determines the signal strength, and the plasma temperature determines the continuum emissions. The above-determined temperature is two times smaller than that obtained in a typical ns-LIBS experiment on lead ($\sim$13, 000 K, averaged without any temporal resolution), using an ArF excimer laser at 193 nm [25], and the measured electron density is similar to those produced in ns-LIBS ($10^{18}\,\text{cm}^{-3}$), typically with the energy of $>$ 50 mJ (see e.g. [26]), which is much larger than that (5 mJ) used in this work. Therefore, the measured high electron density ($\sim 8 \times 10^{17}\,\text{cm}^{-3}$) and low plasma temperature ($\sim$6, 700 K) would give rise to a high signal-to-noise ratio even in the early stage of the plasma, making this technique of FIBS even more appealing.

However, in the filament-induced breakdown spectra, one source that affects the signal-to-noise ratio is the white light produced during the filamentation process in air, as shown in Fig. 7.4. These spectra were recorded with the sample located at (a) 480 cm, (b) 420 cm, (c) 320 cm and (d) 280 cm away from the

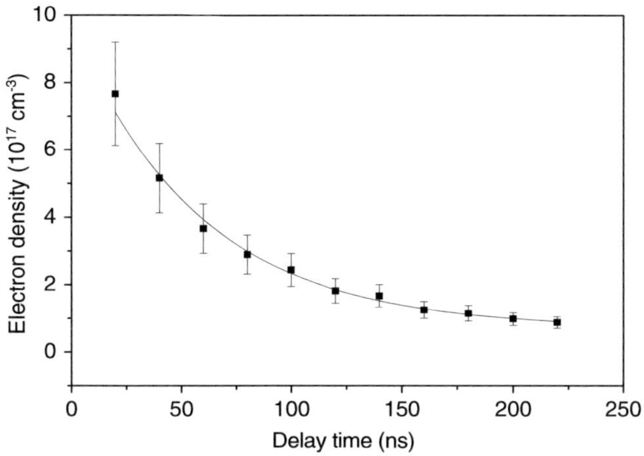

Fig. 7.3 Electron densities of the lead plasma as a function of delay time. The solid line is a least square fit of the experimental data to an exponential decay, showing the behavior of the electron density decay

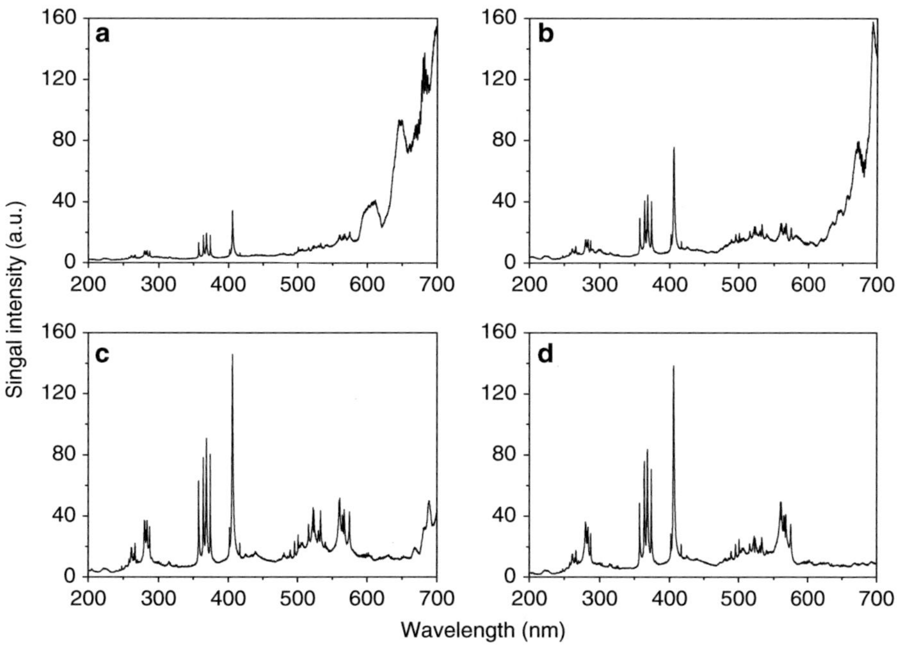

Fig. 7.4 Spectra recorded with a LIDAR configuration. The sample was located at (**a**) 480 cm, (**b**) 420 cm, (**c**) 320 cm and (**d**) 280 cm away from the focusing lens. The ICCD gate widths are set to 2 μs for all the four cases

focusing lens (the filament started at a fixed distance around 2.5 m away from the focusing lens) and a time delay of −3 ns with respect to the laser arriving time on the target. It can be seen from Fig. 7.4 that the white light continuum was much weaker if the sample was placed in the front part (Fig. 7.4d), close to the starting point, of the filament. This can be explained by the fact that the white light spectral width is linearly proportional to the propagation distance [27]. During the filamentation process, self-phase modulation (SPM) occurs inside the filament, leading to a strong spectral broadening of the femtosecond pulse from the ultraviolet (UV) to the infrared (IR) (supercontinuum) [28, 29]. Therefore, when the filament hits the sample surface, the white light in the filament would be scattered back to the detection system. This backscattered white light, however, can be easily eliminated by well controlling the interaction position of the filament with the sample. In particular, when compared to conventional ns-LIBS [30,31], the continuum emission associated with the plasma itself, i.e., free–free or free–bound transitions that result from collisions between electrons and ambient gas species and electron–ion recombination in the plasma, is much weaker because of the short pulse duration and the relatively low plasma temperature. This property of FIBS is very favorable for practical, remote applications because the detection window can be opened much earlier than in classical ns-LIBS, making even nongating technique feasible [32].

7.4 Physical Properties of the Plasma Filament in Air

The characterization of the long plasma filament by the optical emission spectroscopy was also demonstrated in air [33]. Figure 7.5 shows a typical spectrum of the atomic oxygen line emissions in the spectral range of 776.5–778.5 nm. The

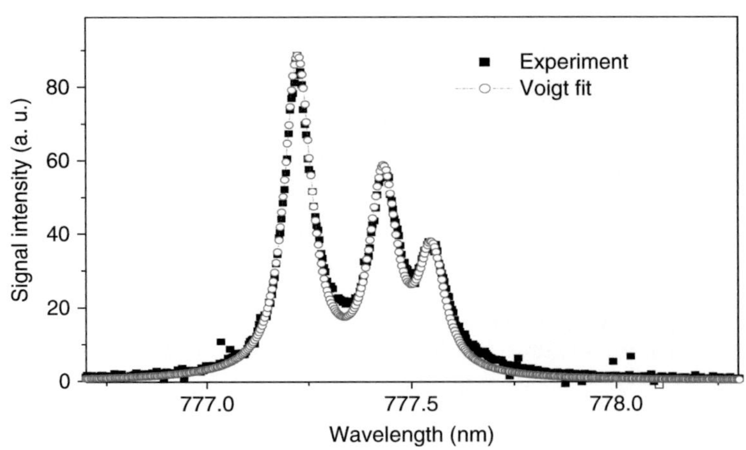

Fig. 7.5 The spectrum of atomic oxygen lines in the plasma filament

spectral lines at 777.19 nm, 777.42 nm, and 777.54 nm were assigned to the atomic transitions $2s^22p^3(^4S^0)3s$–$2s^22p^3(^4S^0)3p$, where $J_k = 3, 2$, and 1, respectively [34]. The intensity ratio of the individual peaks is close to 7:5:3, which is given by the ratio of the statistical weights of the upper levels of the transitions [35]. In this measurement, the laser pulses were focused in the ambient air to generate a single filament by a lens with $f = 100$ cm. The laser pulse had an energy of 2 mJ and a pulse duration of about 42 fs. The pulse spectrum was centered at 805 nm. The spectrum was recorded with an ICCD gate width of 20 ns and a delay time of 0 ns with respect to the laser arrival.

To determine the electron density, the contributions from the collision with ions represented by the second term of (7.4) were first estimated with the values of $A = 0.035$ [18], $\omega = 0.0166$ nm [33], and the plasma parameters of $N_e \sim 10^{16}\,\text{cm}^{-3}$ and $T = 5,800$ K (see later). The broadening ascribable to the collision with ions was only $\sim$ 0.01 nm, which is much less than the measured spectral line widths ($\sim$0.1–0.2 nm). Therefore, the second term of (7.4) was negligible, and (7.5) was used to directly determine the plasma density with the value of $\omega = 0.0166$ nm adopted. As an example, the spectral lines in Fig. 7.5 were fitted well to the convolutions of a Lorentzian profile corresponding to the Stark broadening and a Gaussian profile corresponding to the instrumental broadening (0.03 nm, fixed).

In Fig. 7.6a, the longitudinal electron density distribution of femtosecond laser plasma filament is shown. It can be seen that the electron densities are in the order of $10^{16}\,\text{cm}^{-3}$, which were good agreement with the measurement by other methods [36]. The plasma density of $\sim 2.2 \times 10^{16}\,\text{cm}^{-3}$ was about two orders of magnitude higher than the critical density of $N_{cr} \sim 10^{14}\,\text{cm}^{-3}$ for LTE, when the energy difference ($\Delta E = 1.59$ eV) of the atomic oxygen triplet and the plasma temperature of 5,800 K (see later) were adopted in (7.1). This justifies the condition of LTE.

The electron density recorded in Fig. 7.6a was over a range of about 7 cm and the plasma column was stable within a region of about 4 cm in the case of the

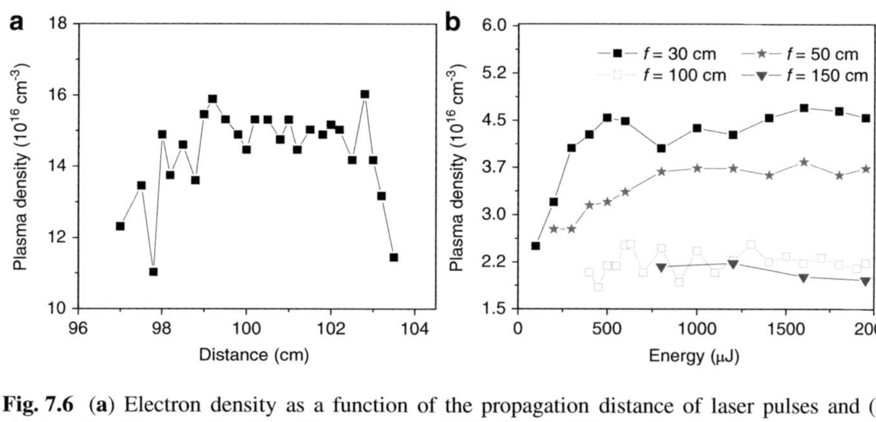

Fig. 7.6 (**a**) Electron density as a function of the propagation distance of laser pulses and (**b**) electron densities as a function of laser energy obtained with different focusing lenses

$f = 100$-cm lens. This behavior was explained by intensity clamping [37, 38], and the observed self-stabilization is fundamental to the filamentation process [39]. The measurements of the electron densities as a function of the laser energy under different focusing conditions were also performed, as shown in Fig. 7.6b. In the case of the shorter focal lengths ($f = 30$ and 50 cm), increasing the input laser energy, the plasma density increases rapidly until the slope tends toward a constant at high laser energy. This latter effect is again due to intensity clamping, which leads to saturation of the laser intensity and therefore the plasma density. The characteristic energy of about 0.5 mJ, indicated by the change of the slope in the plasma density, corresponds to a power of about 10 GW, which is the critical power for self-focusing measured in air [40]. The steep rise below the critical power results from the highly nonlinear dependence of the plasma density on the laser intensity before intensity clamping occurs [36]. This can be observed for the focal lengths of $f = 30$ and 50 cm. In the case of the $f = 100$ and 150-cm lens, the signal-to-noise ratio below the critical power was not sufficient to detect. The increase of the electron density with shorter focal length agrees with earlier measurements using nitrogen fluorescence calibrated by longitudinal diffraction [36].

The plasma temperature was determined using two argon lines at 696.54 and 430.01 nm by mixing 25% argon and 75% air at atmospheric pressure. It was assumed that the plasma temperature in the mixture was not significantly different from that in pure air. According to (7.3), the plasma temperature was estimated to be 5,819 K, which was averaged over the whole length of the plasma filament. No significant change of the plasma temperature was observed along the plasma column. This plasma temperature is relatively low compared to that in a typical nanosecond laser-induced air plasma, i.e., $2\text{–}3 \times 10^4$ K, measured within ≤ 100 ns after the laser arrival [35]. This plasma temperature in the order of $\sim 5 \times 10^3$ K, together with the relatively small electron density of $\sim 10^{16}\,\text{cm}^{-3}$, gives rise to a good signal-to-noise ratio signal.

7.5 Lasing Actions Occurring in the Plasma Filament in Air

An interesting phenomenon, i.e., amplified spontaneous emissions (ASE) along the direction of the plasma filament that could be important in applications such as remote sensing, was observed in 2003 by detecting the backscattering nitrogen fluorescence [41]. The experiments were carried out by using a femtosecond laser system characterized with a pulse duration of 42 fs and an energy of up to 20 mJ at a repetition rate of 10 Hz (Fig. 7.7a). The laser pulses were focused in air by a 100 cm lens to generate a single plasma filament with a length of a few centimeters that was determined by the input laser energy. The nitrogen fluorescence was observed in the backward direction. With such conditions, the backscattered fluorescence intensity of N_2 at 357 nm was found to increase exponentially with a gain coefficient of $0.3\,\text{cm}^{-1}$, as the plasma filament length increases (Fig. 7.7b) Other fluorescence lines of the second positive band from nitrogen molecules showed similar gain.

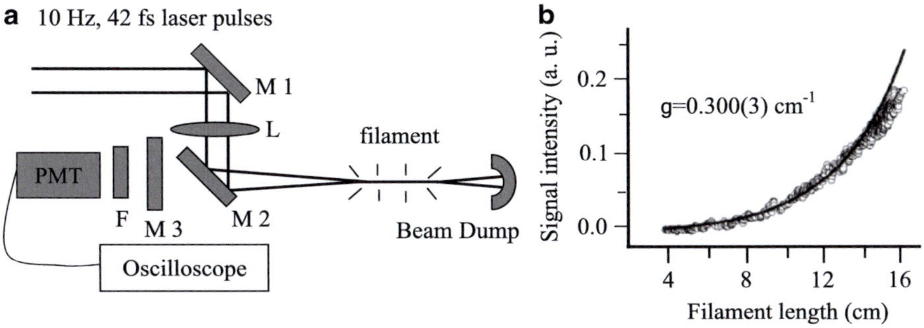

Fig. 7.7 (**a**) The experimental setup. L: focusing lens with $f = 100\,\mathrm{cm}$; M1, M2: dielectric mirrors for 800 nm for 45 degree incident angle; M3: dielectric mirror for 800 nm for 0 degree incidence angle; F: interference filter. The distance between M2 and the beam dump is around 2 m. (**b**) The experimental data (*dots*) and the fitted gain curve (*solid line*)

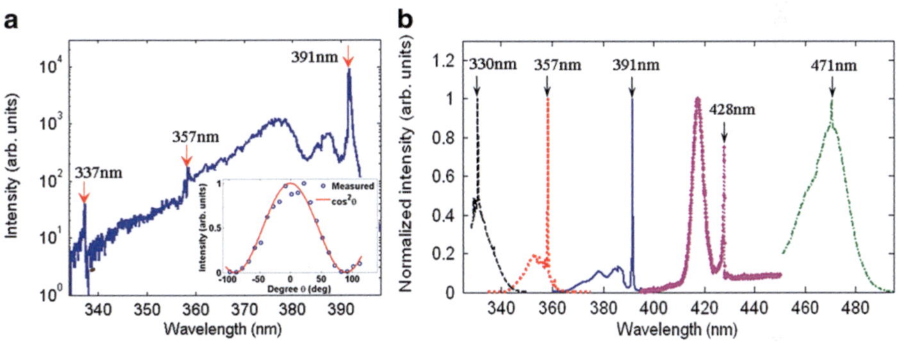

Fig. 7.8 (**a**) Typical spectrum of nitrogen molecules excited by 1900 nm laser pulses recorded in forward directions. Inset in (**a**): polarization property of remote laser at ~391 nm. (**b**) the laser peaks at 471, 428, 391, 357, and 330 nm achieved with different pump wavelengths of 1415, 2050, 1920, 1760, and 1682 nm, respectively.

In addition, the fluorescence signal of N_2^+ at 391 nm was measured, by which a gain coefficient of $0.34\,\mathrm{cm}^{-1}$ was obtained. The mechanisms of the population inversion for the ASE emissions are not totally clear yet, but they were tentatively ascribed to the recombination of free electrons with ions in the plasma. Similar ASE lasing action in air was also demonstrated for atomic oxygen line at 845 nm pumped by a 226-nm, 100-ps laser pulses by a group in the USA [42]. In this circumstance, population inversion for the $3p^3P$–$3s^3S$ transition was ascribed to resonant two-photon excitation of atomic oxygen fragment in the plasma.

Recently, we demonstrated a totally different remote lasing scheme in air, which achieved a self-harmonic-seeded switchable multiwavelength laser in air driven by intense mid-infrared femtosecond laser pulses [43]. As shown in Fig. 7.8a, a strong, narrow-bandwidth emission at 391 nm appears on top of the spectrum of the 5th harmonic, whose intensity is nearly $2 \sim 3$ orders of magnitude higher than that of the fluorescence lines at 357 nm and 337 nm from neutral molecular nitrogen. The forward-direction spectrum was obtained with the laser pulses at 1,900 nm, 500 μJ, and 200 fs. The lasing action was confirmed by the measurement of the polarization of the emission at 391 nm using a Glan-Taylor polarizer. As shown in the inset of Fig. 7.8a, the strong emission at 391 nm is nearly perfectly linearly polarized in the direction parallel to that of the pump pulse. This provides strong evidence on the harmonic-seeded lasing action at 391 nm because both spontaneous emission and ASE will show isotropic polarization. Moreover, the seeding effect cannot take place in the side and backward directions as the 5th harmonic beam always follows the laser propagation in the forward direction; the backward and side emissions spectra do not show any lasing peak as the 391 nm and will appear similar to each other with comparable fluorescence line intensities [44]. This is indeed what we have observed. We also examined the gain of the line at 391 nm, and as a result a gain coefficient of 5.02 cm^{-1} was obtained. This value is about one order of magnitude larger than those obtained in the ASE cases.

By tuning the pump mid-infrared femtosecond laser pulses, the lasing actions at different wavelengths were obtained, as shown in Fig. 7.8b. All the lasing wavelengths were assigned to the first negative band of $N_2{}^+$. The population inversion occurring at an ultrafast time-scale in the multi-wavelength remote lasing actions was ascribed to direct formation of excited molecular nitrogen ions by strong-field ionization of inner-valence electrons, which is fundamentally different from the previously reported pumping mechanisms based either on electron recombination of ionized molecular nitrogen or on resonant two-photon excitation of atomic oxygen fragments resulting from resonant two-photon dissociation of molecular oxygen.

7.6 Application to Atmospheric Sensing

In this section, we will give several representative examples carried out in recent studies for sensing the samples in different phases including gases, vapors, aerosols and solids using filament-induced nonlinear spectroscopy [4, 13].

For gas, vapor and aerosol samples in air, the filamentation of femtosecond laser pulses would induce fragmentation in all these matters in air, resulting in the characteristic fluorescence emissions from the excited fragments along the plasma filament. In Fig. 7.9, a backward fluorescence spectrum of methane (CH_4) mixed with air at atmospheric pressure was demonstrated [45]. In this measurement, the concentration of CH_4 was 26% (volume/volume). The laser pulses with an energy of 5 mJ were focused into a 4.2-m-long tube by a BK 7 lens ($f = 1$ m). The fluorescence from dissociated CH radicals was used to quantitatively analyze the concentration of CH_4 and its remote detection limit. The dissociation mechanism of CH_4 was theoretically ascribed to the neutral dissociation through superexcited states created by multiphoton/tunnel excitation [46]. So far, the fluorescence from the gas samples such as ethanol vapor and smoke (from burning mosquito coils

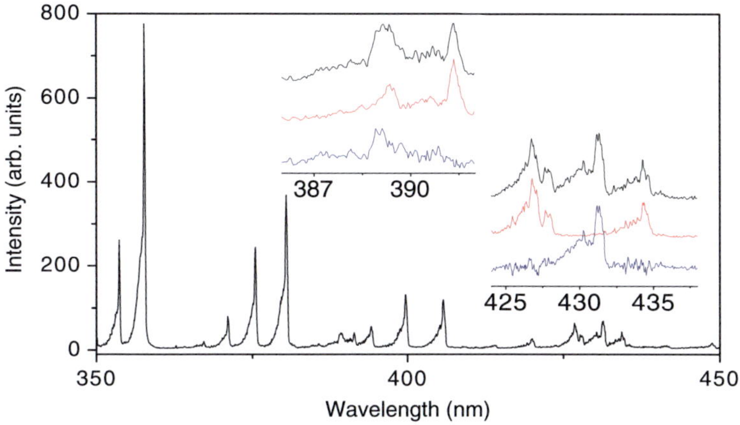

Fig. 7.9 Filament-induced fluorescence spectrum of mixture of CH_4 and air with a CH_4 concentration of 2.6%. The insets show the spectra in a higher resolution (*top*), the spectra of pure air in atmospheric pressure (*middle*), and the subtraction of the mixture and pure air spectra (*bottom*). The bands are CH at 389 nm (B-X) and 431 nm (A-X)

in air) as well as water aerosols containing multiple solutes (NaCl, $PbCl_2$, $CuCl_2$ and $FeCl_2$) [47–50] was recorded. In particular, a simultaneous detection and identification of methane and acetylene mixed with air at atmospheric pressure were also performed using the filament-induced nonlinear spectroscopy [51]. A genetic algorithm has been used to identify the unknown spectra with the premise that a spectral database including the spectral signatures and the strengths of the signals of the corresponding trace species is built. A good agreement between calculation and experiment for the observed species was obtained (Fig. 7.10). Concentration measurements demonstrate that the detection limits could reach about 1 ppm and 300 ppb for CH_4 and C_2H_2, respectively. Very interestingly, we found recently an exponential growth of the backscattered fluorescence from CH radicals emitted from the mixture of hydrocarbon molecules with air at atmospheric pressure (i.e., 2% methane, acetylene, or ethylene) as a function of filament length, indicating that the fluorescence has been amplified through ASE as it propagates along the filament [52].

For the detection of solid samples, we have experimentally shown the remarkably distinct spectra of egg white and yeast powders using time-resolved FIBS [53]. In particular, we demonstrated the feasibility of remote detection and differentiation of some very similar agriculture-related bioaerosols, namely barley, corn, and wheat grain dusts, using this technique [54]. The signals were detected in Lidar configuration. All the species showed identical spectra, namely, those from molecular C_2 and CN bands, as well as atomic Si, C, Mg, Al, Na, Ca, Mn, Fe, Sr, and K lines. These identical spectral bands and lines reveal similar chemical compositions; however, the relative intensities of the spectra are different showing different element abundances from these three biotargets. The intensity ratios of

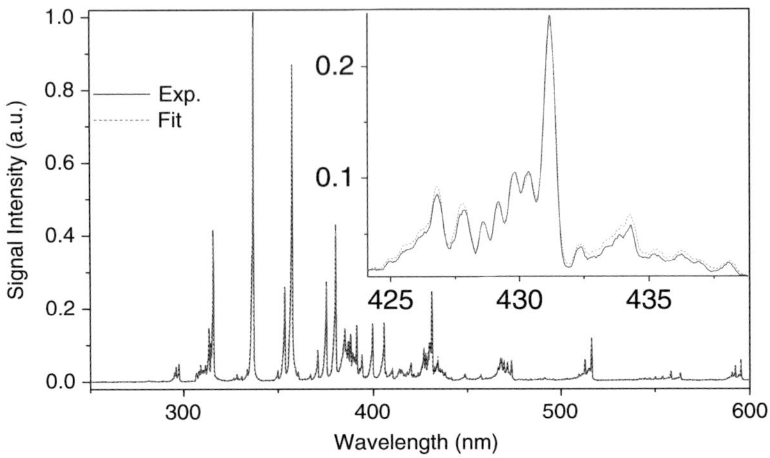

Fig. 7.10 Experimental and fitted filament-induced fluorescence spectra of a mixture of C_2H_2 (1316 ppm), CH_4 (5263 ppm) and air at atmospheric pressure. The inset shows the experimental and fitted spectra in a higher resolution

different elemental lines were used to distinguish these three samples. We have also demonstrated that the usage of a simple telescope as sending optics (see [55]) could greatly improve the performance of remote FIBS of metallic targets (aluminum) [32]. In this case, because the filaments are short, white light continuum inside remote FIBS spectrum is negligible, realizing nongated R-FIBS.

7.7 Summary

In this chapter, we gave an analysis of plasma characteristics of different phase targets when interacting with femtosecond laser filaments by using optical emission spectroscopy. By measuring the widths and strengths of spectral lines, we determined the electron temperature and plasma density of the plasma plume of lead and the plasma filament of air induced by femtosecond laser filamentation. It was shown that the plasma spectra are very clean with a relatively high signal-to-noise ratio because of the low plasma temperature and high plasma density. We also presented the lasing actions occurring in the plasma column in air and discussed the feasibility of remote sensing of pollutants in the atmosphere by the demonstrations of several experimental examples based on filament-induced nonlinear spectroscopy and lasing behaviors.

Acknowledgements We acknowledge the technical support of M. Martin. This work was partially supported by NSERC, DRDC Valcartier, Canada Research Chairs, CIPI, CFI, Femtotech and FQRNT, National Basic Research Program of China (Grant 2011CB808102), National Natural

Science Foundation of China (Grant Nos. 10974213, 60825406, 11074098), NCET-09–0429, and the basic research program of Jilin University.

References

1. S. Svanberg, *Differential absorption Lidar (DIAL)*, in ed. by M.W. Sigrist. *Air Monitoring by Spectroscopic Techniques*, Chap. 3 (Wiley, New York, 1994)
2. S.L. Chin, *Femtosecond Laser Filamentation* (Springer-Verlag, New York, 2010)
3. J. Kasparian, J.-P. Wolf, Opt. Express **16**, 466 (2008)
4. S.L. Chin, H.L. Xu, Q. Luo, F. Théberge, W. Liu, J.F. Daigle, Y. Kamali, P.T. Simard, J. Bernhardt, S.A. Hosseini, M. Sharifi, G. Méjean, A. Azarm, C. Marceau, O. Kosareva, V.P. Kandidov, N. Aközbek, A. Becker, G. Roy, P. Mathieu, J.R. Simard, M. Châteauneuf, J. Dubois, Appl. Phys. B **95**, 1 (2009)
5. S.L. Chin, S.A. Hosseini, W. Liu, Q. Luo, F. Théberge N. Aközbek A. Becker V.P. Kandidov O.G. Kosareva H. Schroeder Can, J. Phys. **83**, 863 (2005)
6. A. Couairon, A. Mysyrowicz, Phys. Rep. **441**, 47 (2007)
7. L. Berge, S. Skupin, R. Nuter, J. Kasparian, J.-P. Wolf, Rep. Prog. Phys. **70**, 1633 (2007)
8. M. Rodriguez, R. Bourayou, G. Méjean, J. Kasparian, J. Yu, E. Salmon, A. Scholz, B Stecklum, J. Eislöffel, U. Laux, P. Hatzes, R. Sauerbrey, L. Wöste, J.-P. Wolf, Phys. Rev. E **69**, 036607 (2004)
9. G. Méchain, G. Méjean, R. Ackermann, P. Rohwetter, Y.B. André, J. Kasparian, B. Prade, K. Stelmaszczyk, J. Yu, E. Salmon, W. Winn, L.A. Schlie, A. Mysyrowicz, R. Sauerbrey, L. Wöste, J.-P. Wolf, Appl. Phys. B **80**, 785 (2005)
10. S.L. Chin, A. Talebpour, J. Yang, S. Petit, V.P. Kandidov, O.G. Kosareva, M.P. Tamarov, Appl. Phys. B **74**, 67 (2002)
11. R. Ackermann, G. Méjean, J. Kasparian, J. Yu, E. Salmon, J.-P. Wolf, Opt. Lett. **31**, 86 (2006)
12. F. Théberge, N. Aközbek, W. Liu, A. Becker, S.L. Chin, Phys. Rev. Lett. **97**, 023904 (2006)
13. H.L. Xu, S.L. Chin, Sensors **11**, 32 (2011)
14. B. La Fontaine, F. Vidal, Z. Jiang, C.Y. Chien, D. Comtois, A. Desparois, T.W. Johnston, J.C. Kieffer, H. Pépin, H.P. Mercure, Phys. Plasma **6**, 1615 (1999)
15. H. Schillinger, R. Sauerbrey, Appl. Phys. B **68**, 753 (1999)
16. S. Tzortzkis, M.A. Franco, Y.-B. André, A. Chiron, B. Lamouroux, B.S. Prade, A. Mysyrowicz, Phys. Rev. E. **60**, R3505 (1999)
17. J. Liu, Z. Duan, Z. Zeng, X. Xie, Y. Deng, R. Li, Z. Xu, S.L. Chin, Phys. Rev. E **72**, 026412 (2005)
18. H.R. Griem, *Plasma Spectroscopy* (McGraw-Hill, New York, 1964)
19. D.A. Cremers, L.J. Radziemski, *Handbook of Laser-induced Breakdown Spectroscopy* (Wiley, New York, 2006)
20. R.W.P. McWhirter, *Spectral intensities*, in eds. by R.H. Huddlestone, S.L. Leonard. *Plasma diagnostic techniques*, Chap. 5 (Academic, New York, 1965)
21. H.L. Xu, J. Bernhardt, P. Mathieu, G. Roy, S.L. Chin, J. Appl. Phys. **101**, 033124 (2007)
22. D.R. Wood, K.L. Andrew, J. Opt. Soc. Am. **58**, 818 (1968)
23. A. Alonso-Medina, C. Colón, C. Herrán-Martinez, J. Quant. Spectrosc. Radiat. Transfer. **68**, 351 (2001)
24. I.S. Fishman, E.V. Sarandaev, M. Kh. Salakhov, J. Quant. Spectrosc. Radiat. Transfer. **52**, 887 (1994)
25. Y.I. Lee, S.P. Samuel, P. Sawan, T.L. Thiem, Y.Y. Teng, J. Sneddon, Appl. Spectrosc. **46**, 436 (1992)
26. M. Sabsabi, P. Cielo, Appl. Spectrosc. **49**, 499 (1995)
27. F. Théberge, W. Liu, S.A. Hosseini, Q. Luo, S.M. Sharifi, S.L. Chin, Appl. Phys. B **81**, 131 (2005)

28. O.G. Kosareva, V.P. Kandidov, A. Brodeur, C.Y. Chien, S.L. Chin, Opt. Lett. **22**, 1332 (1997)
29. W. Liu, S. Petit, A. Becker, N. Aközbek, C.M. Bowden, S.L. Chin, Opt. Commun. **202**, 189 (2002)
30. A. De Giacomo, M. Dell'Aglio, A. Santagata, R. Teghil, Spectrochim. Acta Part B **60**, 935 (2005)
31. B. Le Drogoff, J. Margot, M. Chaker, M. Sabsabi, O. Barthélemy, T.W. Johnston, S. Laville, F. Vidal, Y. von Kaenel, Spectrochim. Acta Part B **56**, 987 (2001).
32. W. Liu, H.L. Xu, G. Méjean, Y. Kamali, J.-F. Daigle, A. Azarm, P.T. Simard, P. Mathieu, G. Roy, S.L. Chin, Spectrochim. Acta Part B **62**, 76 (2007).
33. J. Bernhardt, W. Liu, F. Théberge, H.L. Xu, J.F. Daigle, M. Châteauneuf, J. Dubois, S.L. Chin, Opt. Comm. **281**, 1268 (2008)
34. NIST. See http://physics.nist.gov/PhysRefData/ASD/index.html Accessed in 2007
35. L.J. Radziemski, T.R. Loree, D.A. Cremers, N.M. Hoffman, Anal. Chem. **55**, 1246 (1983)
36. F. Théberge, W. Liu, P.T. Simard, A. Becker, S.L. Chin, Phys. Rev. E **74**, 036406 (2006)
37. J. Kasparian, R. Sauerbrey, S.L. Chin, Appl. Phys. B 71, 877 (2000)
38. A. Becker, N. Aközbek, K. Vijayalakshmi, E. Oral, C.M. Bowden, S.L. Chin, Appl. Phys. B **73**, 287 (2001)
39. S.L. Chin, F. Théberge, W. Liu, Appl. Phys. B **86**, 477 (2007)
40. W. Liu, S.L. Chin, Opt. Express **13**, 5750 (2005)
41. Q. Luo, W. Liu, S.L. Chin, Appl. Phys. B **76**, 337 (2003)
42. A. Dogariu, J.B. Michael, M.O. Scully, R.B. Milles, Science **331**, 442 (2011)
43. J.P. Yao, B. Zeng, H.L. Xu, G.H. Li, W. Chu, J.L. Ni, H.S. Zhang, S.L. Chin, Y. Cheng, Z.Z. Xu, Phys. Rev. A **84**, 051802(R) (2011)
44. A. Talebpour, M. Abdel-Fattah, A.D. Bandrauk, S.L. Chin, Laser Phys. **11**, 68 (2001)
45. H.L. Xu, J.F. Daigle, Q. Luo, S.L. Chin, Appl. Phys. B **82**, 655 (2006)
46. F. Kong, Q. Luo, H.L. Xu, M. Sharifi, D. Song, S.L. Chin, J. Chem. Phys. **125**, 133320 (2006)
47. Q. Luo, H.L. Xu, S.A. Hosseini, J.-F. Daigle, F. Theberge, M. Sharifi, S.L. Chin, Appl. Phys. B **82**, 105 (2006)
48. J.-F. Daigle, Y. Kamali, G. Roy, S.L. Chin, Appl. Phys. B **93**, 759 (2008)
49. J.F. Daigle, G. Méjean, W. Liu, F. Théberge, H.L. Xu, Y. Kamali, J. Bernhardt, A. Azarm, Q. Sun, P. Mathieu, G. Roy, J.R. Simard, S.L. Chin, Appl. Phys. B **87**, 749 (2007)
50. J.F. Daigle, P. Mathieu, G. Roy, J.R. Simard, S.L. Chin, Opt. Commun. **278**, 147 (2007)
51. H.L. Xu, Y. Kamali, C. Marceau, P.T. Simard, W. Liu, J. Bernhardt, G. Méjean, P. Mathieu, G. Roy, J.-R. Simard, S.L. Chin, Appl. Phys. Lett. **90**, 101106 (2007)
52. Y. Kamali, A. Azarm, J.-F. Daigle, S. L. Chin submitted (2012)
53. H.L. Xu, W. Liu, S.L. Chin, Opt. Lett. **31**, 1540 (2006)
54. H.L. Xu, G. Méjean, W. Liu, Y. Kamali, J.F. Daigle, A. Azarm, P.T. Simard, P. Mathieu, G. Roy, J.-R. Simard, S.L. Chin, Appl. Phys. B **87**, 151 (2007)
55. W. Liu, F. Theberge, J.-F. Daigle, P.T. Simard, S.M. Sarifi, Y. Kamali, H.L. Xu, S.L. Chin, Appl. Phys. B **85**, 55 (2006)

Chapter 8
Cascaded Laser Wakefield Acceleration Scheme for Monoenergetic High-Energy Electron Beam Generation

Jiansheng Liu, Wentao Wang, Haiyang Lu, Changquan Xia, Mingwei Liu, Wang Cheng, Aihua Deng, Wentao Li, Hui Zhang, Jiancai Xu, Xiaoyan Liang, Yuxin Leng, Xiaoming Lu, Cheng Wang, Jianzhou Wang, Baifei Shen, Kazuhisa Nakajima, Ruxin Li, and Zhizhan Xu

Abstract Cascaded laser wakefield acceleration (LWFA) of electrons is promising for producing monoenergetic electron beams well beyond 1 GeV by separating and controlling electron injection and postacceleration in two LWFA stages. We have demonstrated that electrons with Maxwellian spectra generated from the first LWFA assisted by tunnel-ionization-induced injection were seeded into the second LWFA and then accelerated to be a 0.8 GeV quasi-monoenergetic electron beam. Further acceleration toward multi-GeV may be fulfilled with a long plasma channel. Optical guiding of intense femtosecond laser pulses for powers up to 160 TW over a 4-cm long ablative capillary discharge plasma channel and laser wakefield acceleration of electrons well beyond 1 GeV were experimentally demonstrated. By employing an oxygen-containing ablative capillary, electron beams with energies extending up to 1.8 GeV were generated by using 130 TW, 55 fs laser pulses.

J. Liu · W. Wang · H. Lu · C. Xia · M. Liu · W. Cheng · A. Deng · W. Li · H. Zhang · J. Xu · X. Liang · Y. Leng · X. Lu · C. Wang · J. Wang · B. Shen · R. Li (✉) · Z. Xu (✉)
State Key Laboratory of High Field Laser Physics, Shanghai Institute of Optics and Fine Mechanics (SIOM), Chinese Academy of Sciences (CAS), Shanghai 201800, China
e-mail: michaeljs_liu@mail.siom.ac.cn; wwt1980@yeah.net; hylu@mail.siom.ac.cn; xiachq@mail.ustc.edu.cn; mwliu@siom.ac.cn; wch86@siom.ac.cn; aihuadeng@siom.ac.cn; wentaoli@siom.ac.cn; huizhang007@gmail.com; xujc84@hotmail.com; liangxy@mail.siom.ac.cn; lengyuxin@mail.siom.ac.cn; xiaominglu@siom.ac.cn; wang_cheng@siom.ac.cn; Wjzno127@163.com; bfshen@mail.shcnc.ac.cn; ruxinli@mail.shcnc.ac.cn; zzxu@mail.shcnc.ac.cn

K. Nakajima
State Key Laboratory of High Field Laser Physics, Shanghai Institute of Optics and Fine Mechanics (SIOM), Chinese Academy of Sciences (CAS), Shanghai 201800, China
High Energy Accelerator Research Organization (KEK) 1–1 Oho, Tsukuba, Ibaraki 305–8081, Japan
e-mail: nakajima@post.kek.jp

K. Yamanouchi et al. (eds.), *Progress in Ultrafast Intense Laser Science VIII*, Springer Series in Chemical Physics 103, DOI 10.1007/978-3-642-28726-8_8,
© Springer-Verlag Berlin Heidelberg 2012

8.1 Introduction

Laser wakefield accelerators (LWFAs) offer great potentials to produce high-energy electron beams (e-beams) on a much smaller scale than the conventional radiofrequency accelerators [1–4]. GeV-class monoenergetic e-beams (MEBs) have been generated via a centimeter-scale gas-filled capillary discharge waveguide LWFA [5, 6]. However, the demand of higher energy MEBs could not be fulfilled with the present single-staged LWFA. The cascaded LWFA is promising for generating multi-GeV MEBs, where the electron injection and acceleration can be independently controlled in different stages.

LWFAs via self-injection in the "bubble" regime have a limited energy gain up to 1 GeV due to the required high electron density [7–11]. Experimental results and numerical simulations indicate that the plasma density less than $2 \times 10^{18}\,\mathrm{cm}^{-3}$ is required to produce multi-GeV e-beams: due to that, a long dephasing length can be maintained and the instability due to relativistic self-focusing can be minimized [12]. However, in this low density case, electrons are difficult to be self-trapped into the plasma wake for a controllable acceleration. External injection relying on tunnel ionization is effective for the low electron density LWFA and has been recently demonstrated to generate e-beams with energies well beyond 1 GeV [13]. The advantage of ionization-induced injection is to decrease the required laser intensity and electron densities [14, 15].

In the ionization-induced injection, targets composed of the mixture of helium and a small amount of higher Z gas (such as nitrogen, oxygen and argon) were used. The helium atoms and outer electrons of the high Z atoms are ionized by the leading part of the laser pulse. The ponderomotive force of the laser pulse drives the ionized electrons to form a plasma wake. The inner-shell electrons are ionized at the peak of the laser pulse and slip backward relative to the plasma wake. Some of these electrons acquire enough energy through the longitudinal electric field acceleration and are then trapped. Thus, the critical laser intensity for ionization-induced injection is determined by the threshold for ionizing the inner-shell electrons of high Z atoms. For example, a laser intensity higher than $1.8 \times 10^{19}\,\mathrm{W/cm^2}$, which corresponds to a normalized vector potential of $a = 2.9$, is required to produce O^{7+} ions [13]. However, due to the continuous injection before the pump depletion of driving laser pulses, the accelerated e-beams are far from a mono-energetic distribution [13–15]. If the trapped electrons assisted by the tunnelionization in the first stage can be seeded into the second plasma wave for further acceleration, electron energies can be easily scaled up to multi-GeV with a narrow energy spread.

8.2 A Cascaded Laser Wakefield Accelerator Using Ionization-Induced Injection

The cascaded two-stage LWFA provides a reliable route to obtain multi-GeV MEBs. The ability to stage multi-LWFAs together is very important for eventually implementing 10 GeV and higher energy laser accelerators, whereby the electrons are repeatedly accelerated by the laser wakefields in a manner similar to the conventional accelerators. Previous proof-of-principle staged acceleration experiments have been demonstrated by the external injection of e-beams from a conventional accelerator into a plasma wave [16–18]. A very low acceleration gradient of 2.8 GV/m was obtained. Several injection scenarios have been designed and analyzed for cascaded LWFAs [19]. A staged all-optical LWFA in the self-modulated regime was demonstrated, however, with the e-beam output of continuous spectrum and $\sim$20 MeV maximum energy [20]. Some proposals of two-stage acceleration via density-gradient or tunnel-ionization injection have been suggested [11, 15, 21, 22] but not yet been demonstrated experimentally until the very recent breakthroughs [23, 24].

Here we report on the generation of near-GeV, quasi-monoenergetic e-beams from an all-optical cascaded LWFA. The cascaded LWFA has two LWFAs where two gas cells (GCs) were employed, the first GC filled with a He/O_2 mixture in the first LWFA while the second one filled with only pure He gas in the second one. The first LWFA was used to generate high-energy electron source due to tunnel-ionization-induced injection. The generated electrons with a Maxwellian spectrum (almost 100% energy spread) were then seeded into the second low-density LWFA. The second LWFA with 3-mm-thick GC was used to accelerate the seeded electrons from the first LWFA to be a collimated quasi-monoenergetic e-beam with energy of 0.8 GeV, corresponding to an acceleration gradient of 187 GV/m. As a result of the two GCs, the separation of the electron injection and acceleration was realized experimentally.

8.2.1 *Experimental Setup*

To realize a cascaded LWFA, we separated the electron injector and accelerator by using two GCs filled with He/O_2 mixture and pure He gas flows, respectively. The schematic of the experimental setup for the all-optical cascaded LWFA assisted by tunnel-ionization-induced injection is shown in Fig. 8.1. The first GC was 1-mm thick and consisted of 6% O_2 gas and 94% He gas; the second GC with variable thickness from 1 to 3 mm was filled with pure He gas. The separation between the walls of the two cells was 0.5 mm. Each GC was machined with an outlet, which could minimize the counterflow of the forgoing gas flow. The inset in Fig. 8.1 shows the detailed structure of the two GCs. The experiments were performed on the Femtosecond Petawatt Laser Facility [25]. The 40-fs, 800-nm

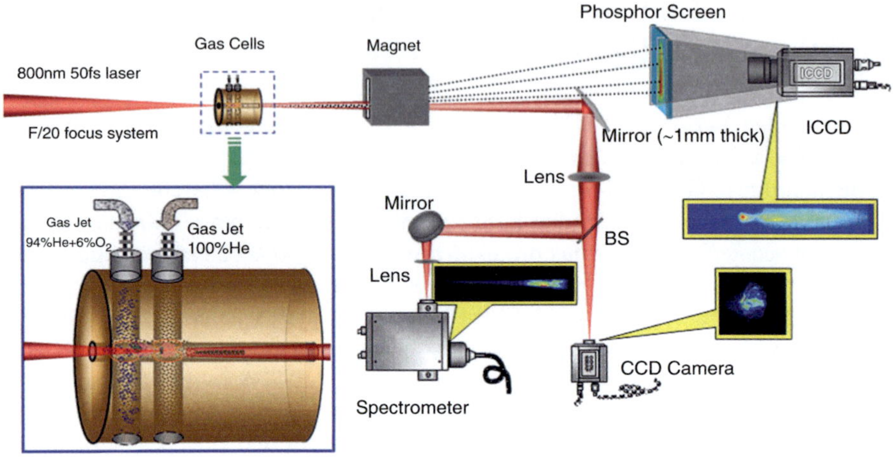

Fig. 8.1 Schematic of experimental setup for the all-optical cascaded LWFA assisted by tunnel-ionization-induced injection. The laser was focused using an f/20 off-axis parabola onto the target, which consisted of two GCs filled with He/O2 mixture and pure He gas flow by using two pulsed gas jets, respectively. The first GC consisted of a mixture of 6% O2 gas and 94% He gas and the thickness was 1 mm. The thickness of the second cell was variable from 1 mm to 3 mm, respectively

laser pulses with powers up to 60 TW were focused by an $f/20$ off-axis parabola into the GC with a beam radius $w_0 \approx 16\,\mu$m at $1/e^2$ of the peak intensity. The produced e-beams were deflected by a 0.55 Tesla dipole magnet and measured by a Lanex phosphor screen imaged onto an intensified charge-coupled device (ICCD), which was cross-calibrated by using a calibrated imaging plate to measure the charge of the e-beams [26]. A 100-μm-thick glass plate was used to send the transmitted light from the GCs to the forward imaging system for alignment and to a spectrometer for transmitted laser spectra measurements. The plasma densities were $\sim 5.7 \times 10^{18}\,\text{cm}^{-3}$ in the first GC and $\sim 2.5 \times 10^{18}\,\text{cm}^{-3}$ in the second one, respectively.

8.2.2 Quasi-Monoenergetic e-Beam Generation from the Cascaded LWFA

Figure 8.2 shows the single-shot e-beam spectra from the first LWFA at different focal positions with laser powers of 50–60 TW (peak intensities of $0.93\text{–}1.1 \times 10^{19}\,\text{Wcm}^{-2}$). The backing pressure of the gas jet was 0.7 bar, corresponding to an electron density of, $n_e \sim 5.7 \times 10^{18}\,\text{cm}^{-3}$ which was measured via a Michelson-type interferometer [27]. As the focal spot was moved from $z = 0$ (the front wall of the first GC) to $z = 1.2$ mm, the cutoff energies of e-beams with Maxwellian spectra

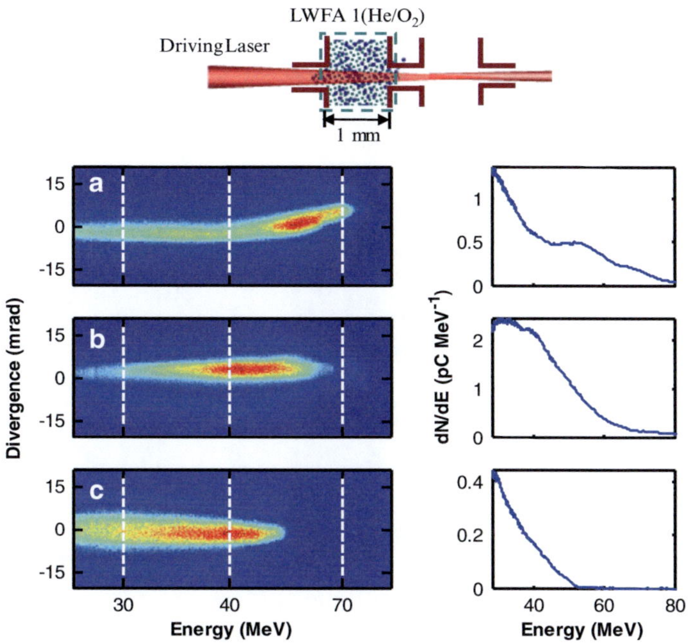

Fig. 8.2 Single-shot e-beam spectra from the first LWFA. Raw e-beam spectra (*left*) recorded by the phosphor screen and an ICCD camera system, and spectra in units of charge per relative energy spread (*right*) at different focal positions: (**a**) The focal position $z = 0$ (the front wall of the first GC) and the laser power on the target was 53 TW. (**b**) $z = 0.6$ mm, 58 TW. (**c**) $z = 1.2$ mm, 60 TW. The electron density was $\sim 5.7 \times 10^{18}$ cm^{-3}. Limited by the exit size of the magnet, only e-beams with energy higher than 27.5 MeV were measured. The generated e-beams have Maxwellian spectra (almost 100% energy spread)

(almost 100% energy spread) decreased from 80 MeV to 50 MeV, and the beam divergence increased slightly from 5 mrad to 8 mrad. The total charges (TCs) of the e-beams with energy >27.5 MeV were 25 pC, 56 pC, and 4.6 pC for Fig. 8.2a–c, respectively. When the first GC was filled with pure He gas, no matter with or without the second one, no electron was observed at backing pressures <1.2 bar ($n_e \sim 7.4 \times 10^{18}$cm^{-3}). As a result, the possibilities of self-injection and density-gradient injection were excluded. We inferred that the ionization-induced injection occurred after the effective laser intensity increased beyond 1.8×10^{19} Wcm^{-2} for the generation of O^{+7} by tunnel ionization [12] due to the self-focusing of the laser beam inside the gas mixture. As the focal spot was moved deeper into the GC, the ionization-induced injection would be postponed, which led to the difference in the TC and the cutoff energy of the e-beams. The first LWFA was used as an electron injector for seeding the second LWFA.

Taking into account the measured Rayleigh range of the focused laser beam as short as ~1 mm, we focused the laser beam at the position $z = 1.2$ mm in order to realize a good self-guiding of laser pulses in the second LWFA and the matching

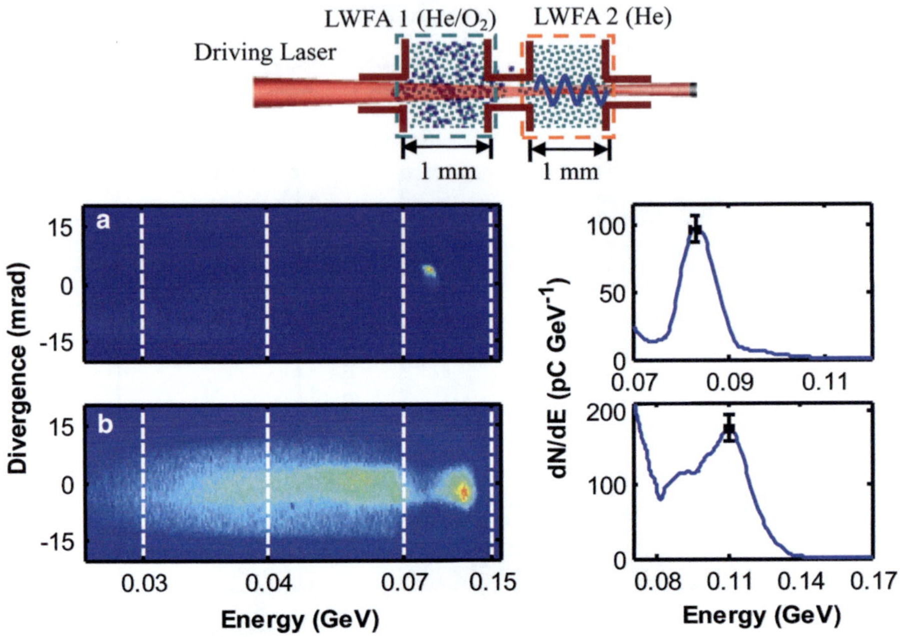

Fig. 8.3 Single-shot e-beam spectra from the cascaded LWFA with 1-mm-thick second GC. Raw electron spectra (*left*) and spectra in units of charge per relative energy spread (*right*) at different laser powers on the target: (**a**) 48 TW and (**b**) 60 TW. The focal position was $z = 1.2$ mm. The electron density was $\sim 2.5 \times 10^{18}\,\mathrm{cm}^{-3}$ in the second gas flow. The vertical error bar ($\pm 11\%$) arose from the uncertainty in the response of the phosphor screen to electrons. The horizontal error bar was owing to the uncertainty in entrance angle of e-beams along the energy dispersion axis. For the 83 MeV (110 MeV) electron beam, this gave an uncertainty in peak energy of $\pm 1.2\%$ ($\pm 1.6\%$)

between the laser and the second plasma wave [28]. Figure 8.3 shows the single-shot e-beam spectra from the cascaded LWFA with 1-mm-thick second GC at different laser powers. The backing pressure of the second gas jet was 0.5 bar, corresponding to $n_e \sim 2.5 \times 10^{18}\,\mathrm{cm}^{-3}$. As shown in Fig. 8.3a, b, e-beams with peak energy of 83 (110) MeV, 9% (27%) energy spread, and 2.6 (7.5) mrad divergence were generated by using 48 (60) TW laser pulses, respectively. The quasi-monoenergetic e-beam at 110 MeV, which was more than doubled of the cutoff energy as shown in Fig. 8.2c, had a charge of $\sim$5.6 pC. It was higher than the measured TC of 4.6 pC obtained in the first LWFA. The TC from the cascaded LWFA was $\sim$87 pC in this case. This result indicated that a great amount of trapped electrons with energies $<$27.5 MeV in the first LWFA were seeded into the second one for further acceleration. The e-beams with almost 100% energy spread were accelerated to be collimated e-beams by the second LWFA. The maximum acceleration gradient of the second LWFA was estimated to be $\sim$100 GV/m.

In the experiments, no electron with energy >27.5 MeV was observed by focusing the laser beam at $z = 1.2$ mm when the first gas jet was off or filled with pure He gas at the same electron density as that of He/O_2 mixture. It is reasonable since the electron density used in the second GC was not high enough for electron self-injection [9, 10]. It was confirmed that the observed high-energy, quasi-monoenergetic e-beams were owing to the post-acceleration in the second LWFA of the e-beams generated via the ionization-induced injection in the first LWFA. Electron injection and efficient acceleration were separately realized.

8.2.3 GeV-Class Quasi-Monoenergetic e-Beam Generation from the Cascaded LWFA with 3-mm-Thick Second Gas Cell

By increasing the acceleration length in the second LWFA larger energy gains were obtained. Figure 8.4 shows single-shot e-beam spectra from the cascaded LWFA with the thickness of the second GC increased to 3 mm. The plasma densities in the first GC and in the second one were $\sim 5.7 \times 10^{18}$ cm^{-3} and $\sim 2.5 \times 10^{18}$ cm^{-3} respectively. By using 50 (48) TW laser pulses, respectively, the e-beams with peak energy of 185 (216) MeV, 20% (19%) energy spread, 4.4 (6) mrad divergence, and $\sim$10 (23) pC charge were generated shown in Fig. 8.4a, b. The maximum energy of the e-beams extended up to $\sim$0.9 GeV as shown in the insets of Fig. 8.4a, b. However, as shown in Fig. 8.4c, a collimated quasi-monoenergetic e-beam peaked at 0.8 GeV was observed with 25% energy spread, 2.6 mrad divergence, and $\sim$3.7 pC charge at 45 TW laser power, in addition to the generation of the e-beam with continuum spectrum around 200 MeV. The energy spread at 0.8 GeV might actually be less (a few percent) as the energy resolution was limited to 25%. Since the uncertainty in the entrance angle of e-beams along the energy dispersion axis was limited to be ± 3 mrad, the resulted uncertainty in peak energy for 0.216 GeV (0.8 GeV) e-beams was estimated to be +4.5%, −3.1% (+19.1%, −10.7%). Considering an effective acceleration length of $\sim$4 mm, the highest acceleration gradient in the second LWFA was estimated to be 187 GV/m. It was close to 70% of 262 GV/m estimated by Lu's Model [27]: $E_{\max} = E_0\sqrt{a_0}$ with $E_0(\mathrm{V/m}) = 96\sqrt{n_e(\mathrm{cm}^{-3})}$, where a_0 was the normalized vector potential.

In order to obtain the maximum acceleration gradient in the second LWFA, electrons should be seeded close to the bubble base. Because the bubble radius of the second plasma wave was larger owing to a lower electron density, lower-energy electrons were generated in the first LWFA and lagged more behind the laser pulse. These lower-energy electrons were favorable for seeding the second LWFA and postacceleration. This might explain the 0.8 GeV quasi-monoenergetic e-beam generated at a relatively low laser power. Another advantage by focusing

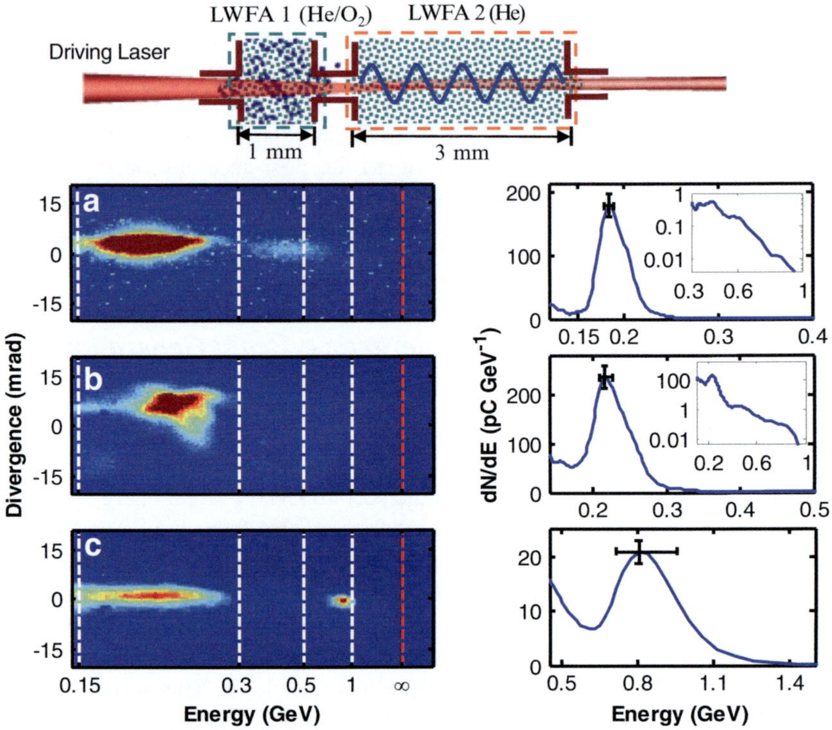

Fig. 8.4 Single-shot e-beam spectra from the cascaded LWFA with 3-mm-thick second GC. Raw electron spectra (*left*) and spectra in units of charge per relative energy spread (*right*) at different laser powers on the target: (**a**) 50 TW, (**b**) 48 TW, and (**c**) 45 TW. The electron densities were $\sim 5.7 \times 10^{18}\,\mathrm{cm}^{-3}$ and $\sim 2.5 \times 10^{18}\,\mathrm{cm}^{-3}$ for the first and second gas flow, respectively. A collimated quasi-monoenergetic e-beam at 0.8 GeV was observed with 25% energy spread, 2.6 mrad divergence and ~3.7 pC charge at 45 TW laser power, in addition to the generation of the e-beam with continuum spectrum around 200 MeV. The energy spread at 0.8 GeV might actually be less (a few percent) as the energy resolution was limited to 25%. For the 0.216 GeV (0.8 GeV) electron beam, the uncertainty in peak energy is +4.5%, −3.1% (+19.1%, −10.7%)

the laser pulse close to the second GC was to produce low-energy e-beams in the first LWFA for the optimization of seeding besides the self-guiding in the second LWFA. Properly controlling and optimizing the injection phase of electrons into the second LWFA by manipulating the laser and the plasma parameters can ensure the stability of e-beams generation. We unambiguously demonstrated for the first time that by using the cascaded LWFA, low-energy e-beams with Maxwellian spectra from an electron injector were postaccelerated to be high-energy, quasi-monoenergetic e-beams. The cascaded LWFA scheme developed here can be scaled up to the generation of 10 GeV e-beams in a straightforward way by using a long preformed plasma channel as the second LWFA.

8.3 Cascaded Laser Wakefield Acceleration of Electron Beams Beyond 1 GeV from an Ablative Capillary Discharge Waveguide

Plasma waveguides preformed by gas-filled or ablative capillary discharges can guide the laser pulse over a much longer distance [5, 7, 29–33] than in the self-guided regime. A quasi-monoenergetic e-beam up to 1 GeV energy was produced by channeling a 40 TW laser pulse in a 3.3-cm-long gas filled capillary discharge waveguide [5]. A meter-scale plasma channel was also proposed to support the LWFA operating at a density of $1.2 \times 10^{17}\,\mathrm{cm}^{-3}$ for generating 10 GeV MEBs by using petawatt-class driving laser pulses [28].

We report a 4-cm-long ablative capillary discharge waveguide cascaded LWFA well beyond 1 GeV assisted by ionization-induced injection, where the capillary had a cascaded structure of 2-cm-long acrylic resin (C:O:H = 4:2:7) capillary connecting with 2-cm-long polyethylene capillary(C:O:H = 1:0:2). By comparing the experiments using polyethylene capillary and the capillary with oxygen-containing acrylic resin at plasma densities of $1\text{–}3\times10^{18}\,\mathrm{cm}^{-3}$, the experimental results suggest that a small amount of oxygen from the ablation of the oxygen-containing capillary wall is in charge of the electron injection, i.e., arising from the ionization of oxygen K-shell electrons. The e-beams with energies up to 1.8 GeV are obtained via the 130 TW, 55 fs driving laser pulses. Optical guiding of 160 TW ultraintense laser pulses is also demonstrated over the entire plasma channel. The results show that it is applicable by using ablative capillary waveguides to produce multi-GeV e-beams. And this cascaded LWFA have proved that ionization-induced injection had some advantages in improving charge of the electron injection, so it could make full use of the acceleration stage to accelerate the e-beams.

8.3.1 Experimental Setup

The experiments were performed at the Femtosecond Petawatt Laser Facility [25]. In the experiments, only 100–160 TW laser pulses were used. The laser pulse duration is 55 fs (full width at half-maximum, FWHM) with the central wavelength at 800 nm. The experimental setup is shown in Fig. 8.5 where the laser beam was focused by an $f/10$ off-axis parabolic mirror into the entrance of an ablative capillary. The vacuum spot size w_0 was measured to be $\sim 13\,\mu\mathrm{m}$ at $1/e^2$ of the peak intensity, corresponding to a Rayleigh length of 0.7 mm. The fractional laser energy contained within the laser spot was measured to be about 55%. The produced e-beams exiting the capillary were deflected by a 10-cm-long 0.55-Tesla dipole magnets and detected by an imaging plate (IP), which was covered by a 100-μm-thick aluminum foil to avoid the exposure from the laser pulse and X-rays. The imaging plate was calibrated to measure the charge of the e-beams [26].

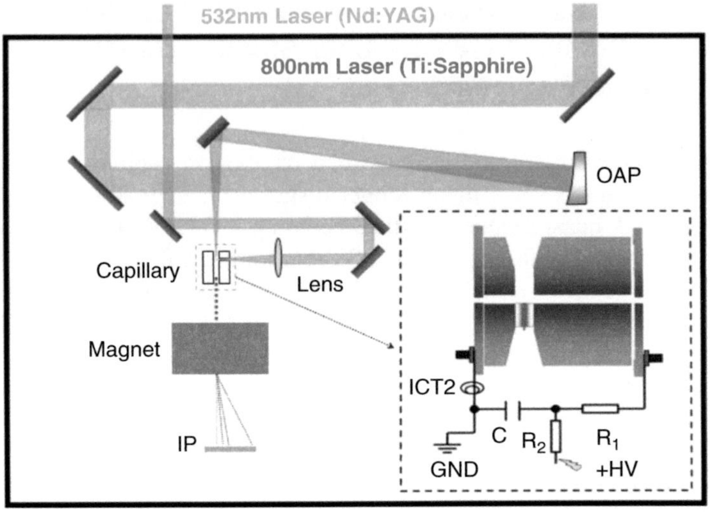

Fig. 8.5 Schematic experimental setup of the ablative capillary discharge waveguide LWFA

A 1-mm-thick glass plate was used to send the transmitted light to the forward imaging system for capillary alignment.

The capillary discharge was triggered several hundred nanoseconds before the driving laser pulse arrived, by focusing a 40 mJ, 532 nm Nd: YAG laser pulse onto the tip of a copper needle with the assistance of the transverse ignition method [33]. This arrangement has an advantage of providing a stable control of the discharge initiation with timing jitters less than 10 ns. The arrangement of the ablative capillary discharge circuit and the diagnostics of the electron densities in terms of the spectral broadening of H_α line were described in details in [33].

8.3.2 Optical Guiding for the Laser from an Ablative Capillary Discharge Waveguide

Optical guiding is important for electron acceleration because the laser intensity will decrease rapidly during the propagation if a proper guiding of laser pulses is not existed. The structure of the capillary had some advantages in ensuring a longer waveguide [29–33].

Figure 8.6a depicts the temporal evolution of the on-axis electron density, discharge current, and the trigger laser pulse of the discharged ablative capillary made of polyethylene. The inner diameter of the capillary was 0.5 mm, and the voltage applied to the capillary electrodes was 20 kV. The maximum on-axis density was increased to $2.5 \times 10^{18}\,\mathrm{cm}^{-3}$ when the time delay relative to the onset of the discharge current was around 410 ns. The measured radial distribution of the

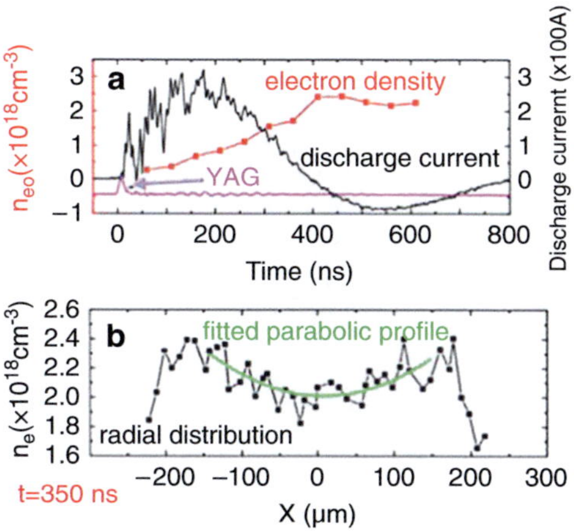

Fig. 8.6 (**a**) Temporal evolution of the on-axis electron density, discharge current, and the trigger laser pulse. (**b**) Radial distribution of the electron densities at a time delay of 350 ns relative to the onset of the discharge current, and the fitted parabolic profile favorable for optical guiding

electron density at the time delay of 350 ns is shown in Fig. 8.6b, with a parabolic profile favorable for optical guiding.

In the first instance, we performed the optical guiding experiments at the Femtosecond Petawatt Laser Facility operated in the 10-Hz mode to confirm the channel structure. The intensity of the driving laser pulse at the focus was $\sim 2 \times 10^{15}$ W/cm^2. Therefore, the relativistic self-focusing effect can be negligible and only the preformed channel contributed to the guiding of the driving laser pulse. Figure 8.7a describes the measured beam patterns of the transmitted laser pulses at the exit when the driving laser pulses were focused into the entrance of the discharge capillary at different time delays from 260 to 440 ns. It can be seen that the spot size of the transmitted laser beam at the exit was approximately equal to the initial spot size at the capillary entrance when the time delay was in the range of 320–380 ns, which indicates the realization of a good linear optical guiding of the laser beam in the plasma channel. This agrees with the measured electron density profile shown in Fig. 8.6b. Furthermore, the nonlinear optical guiding of the driving laser pulses with much higher peak power was also observed at the time delay of 350 ns, as shown in Fig. 8.7b, c, for which the input laser peak powers are 160 TW and 78 TW, respectively. As a result, the nonlinear wakefield for electron acceleration over the entire capillary can be excited by using a 100 TW-level laser pulses.

8.3.3 Cascaded Laser Wakefield Acceleration of Electron Beams Beyond 1 GeV from an Ablative Capillary Discharge Waveguide

Because of the low plasma density inside ablative capillary discharge waveguides, it is difficult for the self-injection to take place, and the production of the e-beams is therefore unstable. However, the advantage of low plasma density is that

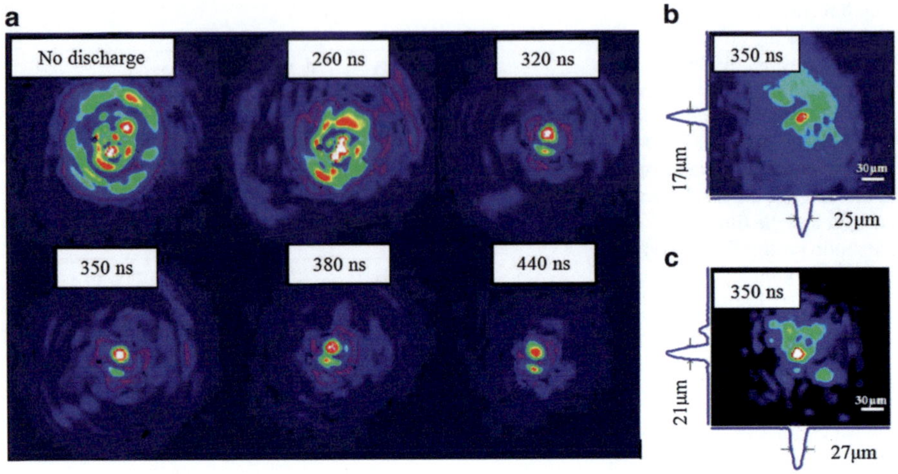

Fig. 8.7 (**a**) Measured beam patterns of the transmitted laser pulses at the exit when the driving laser pulses at intensities of $\sim 2 \times 10^{15}\,\mathrm{W\,cm^{-2}}$ were focused into the entrance of the discharge capillary at different time delays from 260 ns to 440 ns. (**b**)–(**c**) Measured beam patterns of the transmitted laser pulses at the exit at the time delay of 350 ns for input laser powers of 160 TW and 78 TW, respectively

electrons can be accelerated to very high energy after they are properly injected. We performed an experiment for comparison by applying two kinds of capillaries made of different materials. One is 4-cm-long capillary made of polyethylene (C:O:H = 1:0:2), while the other is 2-cm-long acrylic resin (C:O:H = 4:2:7) capillary connecting with 2-cm-long polyethylene capillary.

With the 4-cm-long polyethylene capillary, no electrons were measured when the driving laser powers (on target) varied from 80 to 130 TW and the time delay was scanned in a large range. This is due to the very low electron density in the plasma channel and the self-injection is difficult to occur at this level of driving laser power. This result was quite similar to the experiment reported in [30], where no accelerated electrons were observed for a polyethylene capillary. However, the results were much different for capillaries with oxygen-containing acrylic resin, as shown in Fig. 8.8.

Figure 8.8a, c depict the measured single-shot e-beam energy spectra from an ablative capillary discharge waveguide LWFA. The on-target laser power was 120 TW, and the focal size was estimated to be 21 μm in diameter, corresponding to a peak laser intensity of $1.9 \times 10^{19}\,\mathrm{Wcm^{-2}}$. It was higher than the threshold intensity for the ionization of K-shell electrons of oxygen to produce O^{7+} ions. The time delay between the driving laser pulse and the onset of the discharge current was 410 ns. The discharge voltage applied on the capillary was 23 kV. The corresponding plasma density in the capillary was $2.8 \times 10^{18}\,\mathrm{cm^{-3}}$. It can be seen that the maximum energy of e-beam is up to 1.8 GeV. The uncertainty of the maximum energy at

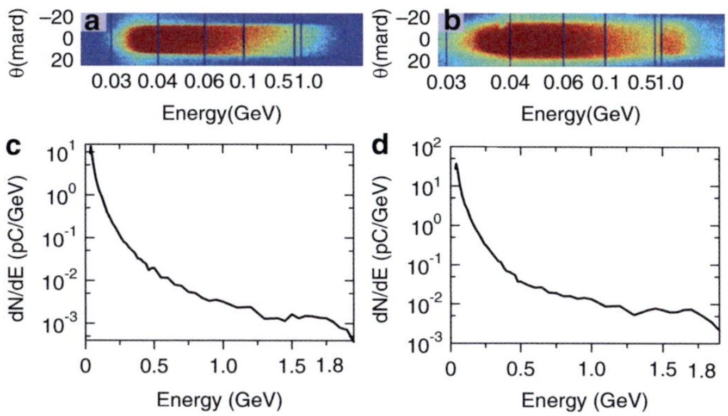

Fig. 8.8 Measured single-shot e-beam energy spectra from an ablative capillary discharge waveguide LWFA at different laser powers on the target. (**a**)–(**b**) Raw data recorded by the imaging plate. (**c**)–(**d**) Integrated spectra in units of charge per relative energy spread. The laser power on the target were 120 TW (*left*) and 130 TW (*right*). The electron densities were $2.8 \times 10^{18}\,\mathrm{cm}^{-3}$ (*left*) and $3.1 \times 10^{18}\,\mathrm{cm}^{-3}$ (*right*). The time delays of the driving laser pulse relative to the onset of the discharge current were 410 ns (*left*) and 490 ns (*right*), respectively

1.8 GeV was estimated to be in the range of 1.2–3.6 GeV considering the divergence angle (4.8 mrad) of the high-energy part of the e-beam. The measured total charge of the e-beam with energies higher than 33 MeV was 0.7 pC.

A higher plasma density of $3.1 \times 10^{18}\,\mathrm{cm}^{-3}$ was achieved by increasing the time delay of the driving laser pulse to 490 ns, and the e-beam with a higher charge of 1.8 pC was observed as shown in Fig. 8.8b, d. The measured maximum energy of generated e-beam also extended up to 1.8 GeV, with a similar uncertainty of 1.2–3.3 GeV (divergence angle 4.5 mrad). The laser power on the target was 130 TW, corresponding to a peak laser intensity of $2 \times 10^{19}\,\mathrm{Wcm}^{-2}$. The increase in the e-beam charge in this case can be partly attributed to the higher plasma density in the capillary waveguide.

The above results suggested that the ionization-induced injection owing to the K-shell electrons of oxygen was responsible for the electron injection into the wakefield [13] and the generation of high-energy MEBs in the case of applying oxygen-containing acrylic resin capillary [31, 32]. And ionization-induced injection had some advantages in improving charge of the electron injection and in ensuring an earlier electron injection. So the cascaded LWFA could make full use of the acceleration stage to accelerate the e-beams. Since the dephasing length at the electron density of $3.1 \times 10^{18}\,\mathrm{cm}^{-3}$ was estimated to be $L_{\mathrm{dp}} = \lambda_{\mathrm{p}}{}^3/\lambda^2 \approx 1.3\,\mathrm{cm}$ [34], which was shorter than the length of the capillary, the generated e-beams had a continuum distribution with 100% energy spread as well as a large divergence angle. Optimizing the capillary length and the inner diameter of the capillary would improve the energy spread and the divergence of the generated e-beams.

8.4 Summary

We have demonstrated an all-optical cascaded laser wakefield accelerator (LWFA) using ionization-induced injection. Electron injection and acceleration were successfully separated and controlled in different LWFA stages. Electrons with Maxwellian spectra generated from the first LWFA using tunnel-ionization-induced injection, were seeded into the second LWFA and accelerated to be a quasi-monoenergetic e-beam with energy of 0.8 GeV, corresponding to an acceleration gradient of 187 GV/m. We have shown the possibility of further acceleration toward multi-GeV by employing a plasma channel. Optical guiding of intense femtosecond laser pulses for powers up to 160 TW over a 4-cm ablative capillary discharge plasma channel and laser wakefield acceleration of electrons well beyond 1 GeV were experimentally demonstrated. By employing an oxygen-containing ablative capillary, e-beams with energies extending up to 1.8 GeV were generated by using 130 TW, 55 fs laser pulses.

A possible geometry for the future experiment is that the second stage of the cascaded LWFA uses the ablative capillary to improve the output electron energy. The ablative capillary discharge plasma channel can offer long acceleration length which can be extended to several centimeters even 10 cm, and the e-beams seeded into the second LWFA can be accelerated to be 10 GeV class mono-energetic e-beams.

Acknowledgements This work was supported by the National 973 Program of China (Contract No: 2011CB808100), NNSF of China (Contract Nos: 10974214, 60921004 and 10834008), the State Key Laboratory Program of Chinese Ministry of Science and Technology, and the Knowledge Innovation Program of CAS. K. Nakajima is supported by CAS visiting professorship for senior international scientists.

References

1. T. Tajima, J.M. Dawson, Phys. Rev. Lett. **43**, 267 (1979)
2. S.P.D. Mangles et al., Nature (Lond.) **431**, 535 (2004)
3. C.G.R. Geddes et al., Nature (Lond.) **431**, 538 (2004)
4. J. Faure et al., Nature (Lond.) **431**, 541 (2004)
5. W.P. Leemans et al., Nat. Phys. **2**, 696 (2006)
6. T.P.A. Ibbotson et al., Phys. Rev. ST Accel. Beams. **13**(031), 301 (2010)
7. S. Karsch et al., New J. Phys. **9**, 415 (2007)
8. N.A.M. Hafz et al., Nat. Photon. **2**, 571 (2008)
9. S. Kneip et al., Phys. Rev. Lett. **103**, 035002 (2009)
10. D.H. Froula et al., Phys. Rev. Lett. **103**, 215006 (2009)
11. J.E. Ralph et al., Phys. Plasma. **17**(056), 709 (2010)
12. S.F. Martins et al., Nat. Phys. **6**, 311 (2010)
13. C.E. Clayton et al., Phys. Rev. Lett. **105**, 105003 (2010)
14. C. McGuffey et al., Phys. Rev. Lett. **104**, 025004 (2010)
15. A. Pak et al., Phys. Rev. Lett. **104**, 025003 (2010)
16. F. Amiranoff et al., Phys. Rev. Lett. **81**, 995 (1998)

17. C.E. Clayton et al., Phys. Rev. Lett. **70**, 37 (1993)
18. M. Everett et al., Nature (Lond.) **368**, 527 (1994)
19. V. Malka et al., Phys. Rev. ST Accel. Beams. **9**(091), 301 (2006)
20. D. Kaganovich et al., Phys. Plasma. **12**(100), 702 (2005)
21. A.G. Khachatryan et al., Phys. Rev. ST Accel. Beams. **10**(121), 301 (2007)
22. R. Trines et al., New J. Phys. **12**(045), 027 (2010)
23. J.S. Liu et al., Phys. Rev. Lett. **107**, 035001 (2011)
24. B.B. Pollock et al., Phys. Rev. Lett. **107**, 045001 (2011)
25. X. Liang et al., Opt. Express. **15**, 335 (2007)
26. K.A. Tanaka et al., Rev. Sci. Instrum. **76**(013), 507 (2005)
27. Z. Zhou et al., J. Phys. B. **43**(135) 603 (2010)
28. W. Lu et al., Phys. Rev. ST Accel. Beams. **10**(061), 301 (2007)
29. K. Nakamura et al., Phys. Plasma. **14**(5), 056708 (2007)
30. C. McGuffey et al., Phys. Plasma. **16**(11), 113105 (2009)
31. T. Kameshima et al., Appl. Phys. Exp. **1**, 066001 (2008)
32. T. Kameshima et al., Phys. Plasma. **16**(9), 093101 (2009)
33. M. Liu et al., Rev. Sci. Instrum. **81**(3), 036107 (2010)
34. E. Esarey et al., Rev. Mod. Phys. **81**(3), 1229 (2009)

Chapter 9
Laser Radiation Pressure Accelerator for Quasi-Monoenergetic Proton Generation and Its Medical Implications

C. S. Liu, X. Shao, T. C. Liu, J. J. Su, M. Q. He, B. Eliasson, V. K. Tripathi, G. Dudnikova, R. Z. Sagdeev, S. Wilks, C. D. Chen, and Z. M. Sheng

Abstract Laser radiation pressure acceleration (RPA) of ultrathin foils of sub-wavelength thickness provides an efficient means of quasi-monoenergetic proton generation. With an optimal foil thickness, the ponderomotive force of the intense short-pulse laser beam pushes the electrons to the edge of the foil, while balancing the electric field due to charge separation. The electron and proton layers form a self-organized plasma double layer and are accelerated by the radiation pressure of the laser, the so-called light sail. However, the Rayleigh–Taylor instability can limit the acceleration and broaden the energy of the proton beam. Two-dimensional particle-in-cell (PIC) simulations have shown that the formation of finger-like structures due to the nonlinear evolution of the Rayleigh–Taylor instability limits

C. S. Liu (✉) · X. Shao · T. C. Liu · J. J. Su · M. Q. He · B. Eliasson · V. K. Tripathi · G. Dudnikova · R. Z. Sagdeev
East-West Space Science Center, University of Maryland, College Park, MD, USA
e-mail: csliu@umd.edu; xshcn@astro.umd.edu; tcliu@umd.edu; jjsu@umd.edu; minqinghe@gmail.com; bengt@tp4.ruhr-uni-bochum.de; vkt@physics.iitd.ernet.in; gdudnikova@gmail.com; rsagdeev@gmail.com

Z. M. Sheng
Department of Physics, Shanghai Jiao Tong University, Shanghai, People's Republic of China

B. Eliasson
Ruhr-University Bochum, D-44780 Bochum, Germany

V. K. Tripathi
Indian Institute of Technology, New Delhi 110016, India

S. Wilks · C. D. Chen
Lawrence Livermore National Laboratory, USA
e-mail: wilks1@llnl.gov; chen41@llnl.gov

Z. M. Sheng
Beijing National Laboratory for Condensed Matter Physics, Institute of Physics, CAS, Beijing, People's Republic of China
e-mail: zmsheng@sjtu.edu.cn

K. Yamanouchi et al. (eds.), *Progress in Ultrafast Intense Laser Science VIII*, Springer Series in Chemical Physics 103, DOI 10.1007/978-3-642-28726-8_9,
© Springer-Verlag Berlin Heidelberg 2012

the acceleration and leads to a leakage of radiation through the target by self-induced transparency. We here review the physics of quasi-monoenergetic proton generation by RPA and recent advances in the studies of energy scaling of RPA, and discuss the RPA of multi-ion and gas targets. The scheme for generating quasi-monoenergetic protons with RPA has the potential of leading to table-top accelerators as sources for producing monoenergetic 50–250 MeV protons. We also discuss potential medical implications, such as particle therapy for cancer treatment, using quasi-monoenergetic proton beams generated from RPA. Compact monoenergetic ion sources also have applications in many other areas such as high-energy particle physics, space electronics radiation testing, and fast ignition in laser fusion.

9.1 Introduction

Compact laser-driven accelerators are an attractive alternative for high-quality monoenergetic proton and ion generation in comparison to conventional RF accelerator because the electric fields for particle acceleration can reach the order of tens GV per cm, which allows reduction of the system size. Laser acceleration of electrons by plasma wakefield was first pointed out by Tajima and Dawson [1], and has become a great success with the experimental production of mono-energetic electrons by laser wakefield acceleration [2–4] and obtaining quasi-monoenergetic electrons close to 1 GeV [5, 6]. More recently, the actively studied scheme of laser radiation pressure acceleration (RPA) of ultrathin target shows promising aspect of efficient quasi-monoenergetic proton generation [7–16]. In the RPA scheme, a circularly polarized, high power, short pulse laser is focused on an ultrathin target (thickness $< \lambda$), which leads to the acceleration of the most of the foil-trapping protons in it, the "light sail." Instead of striving to increase the peak proton energy with a broad energy spectrum using long-pulse laser on a thick target such as the scheme of target normal sheath acceleration (TNSA), the RPA focuses more on increasing the efficiency of acceleration and producing monoenergetic protons.

In this chapter, we review the physics of the laser RPA scheme for generating quasi-monoenergetic protons and carbon ions with potential medical applications such as particle therapy, and present our recent studies of energy scaling of RPA. We also discuss the RPA of multi-ion and gas targets. We focus on the development of laser proton accelerators in the intermediate energy range of few hundred MeV specifically for particle cancer therapy as an immediate goal. First, it is possibly achievable with existing commercially available technology to build a laser RPA-based compact source of high-quality protons (high enough flux, low divergence, nearly monoenergetic and tunable) of energies 50–250 MeV suitable for widespread dissemination for cancer therapy. The potential of compact accelerators can radically enhance the availability of proton therapy device and make it as widespread as MRI machines.

9.1.1 *Brief Introduction on Particle Therapy for Cancer Treatment*

Proton therapy for cancer treatment is a growing industry and rapidly gaining market shares that has been traditionally served by chemotherapy and conventional radiation therapies such as *X-ray* and *gamma ray therapy*. The key feature of the interaction of energetic protons with tissue is the Bragg peak with abrupt absorption and deposition of energy at a specific location and only small amount of energy absorbed elsewhere. This enables precise targeting of the cancer by proton, leading to humane treatment of cancers with minimum damage to the surrounding tissue. Several kinds of cancers can be treated effectively with relatively few side effects for the patients, thus making proton therapy a preferred oncology, particularly in pediatric cases. In proton cancer therapy, protons with energies 50–250 MeV are needed. The proton energy is required to be tunable and nearly monoenergetic so that it can target precisely the tumor location. The particle flux for medical treatment is around 10^9–10^{11} particles per second. Typically, *cyclotron-based proton therapy* centers cost about $150 million and occupy a large real estate. Due to the large size and cost of the current systems, access is limited to a few locations with a capacity of treating only about 6,000 patients per year in the USA, which is less than 3% of the estimated annual demand of 250,000 patients. Though the number of proton therapy centers is increasing, the low-cost compact laser-driven proton accelerators, when realized, would be an attractive alternative for proton cancer therapy.

9.1.2 *Schemes for Laser Proton Acceleration*

The acceleration of ions by laser irradiation of foils has been pursued actively using experiments, theory and simulations. There are mainly two mechanisms for proton acceleration with laser irradiation on a foil: target normal sheath acceleration (TNSA) and radiation pressure acceleration (RPA).

9.1.2.1 Target Normal Sheath Acceleration

In previous studies of the generation of highly energetic ion beams from laser-plasma interactions, foil targets with thicknesses ranging from a few to several tens of laser wavelengths were used and TNSA was the predominant mechanism leading to the production of multi-tens MeV ion beams [17–25]. In TNSA, protons are accelerated by the electric field created by the hot electrons heated by the laser. However, in most cases, the resulting ion energy spectrum is broad and only few protons reach the maximum energy (Fig. 9.1a). In addition, the efficiency of energy conversion to energetic protons is low. Therefore, as pointed out by Ma et al. [34], TNSA is not suitable for direct use in radiotherapy and other applications requiring monoenergetic protons.

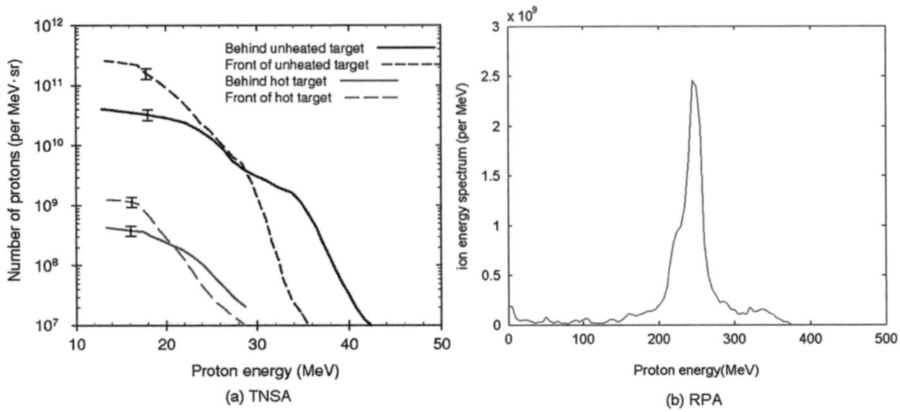

Fig. 9.1 (**a**) Measured proton energy spectra from ultra-high intensity laser interaction with target foils using TNSA scheme (from Ref. [33]). The proton flux from the heated target (*gray lines*), measured at both the front (*broken lines*) and rear (*solid lines*) of the Fe target foil, is reduced by about 2 orders of magnitude compared to the unheated target (*black lines*). (**b**) 2D PIC simulation of proton energy spectra produced with RPA of a thin foil [30]

9.1.2.2 Radiation Pressure Acceleration

In RPA, or "light sail," the whole foil is accelerated by the radiation pressure-trapping protons in it. RPA is a more efficient acceleration process for producing high-energy monoenergetic protons, suitable for many applications requiring that the accelerated protons have a good beam quality and a narrow energy spectrum. We have studied and developed theory and simulation of the RPA of quasi-monoenergetic protons [7, 13, 15, 30]. Figure 9.1b shows the modeled energy spectrum of protons produced from RPA of a thin target. Recent simulations of the RPA scheme focused on case studies of obtaining $\sim$ GeV protons with high power ($\sim$10 of petawatt) laser irradiated on single-species ion target [26, 27] and on the acceleration of minority protons with RPA of multi-species (carbon and proton) thin foil targets [28]. Experiments by [10] demonstrated promising aspects of efficient laser acceleration of ultrathin targets.

9.2 Quasi-Monoenergetic Proton Generation with RPA

9.2.1 Criteria for RPA

In RPA, high-intensity circularly polarized laser light with a high contrast ratio accelerates the ultrathin foil with radiation pressure, and the foil has a definite, optimal thickness. It must be sufficiently thin so that

(a) The ponderomotive force of the laser radiation accelerating the electrons in the foil is balanced by the electric force due to ions left behind at the outer edge of the thin foil.
(b) The mass of the thin foil must be sufficiently light so that nearly the whole foil is accelerated by the laser radiation pressure in the short duration of the order of ion plasma period. In an accelerated frame, protons are subject to both the electric force of the electron layer accelerating them forward, and the inertial force pulling them back. The balance of these two opposing forces forms a trap for the ions in real and phase spaces.
(c) These stably trapped protons and electron layers form a self-organized double layer. The laser radiation pressure accelerates this double layer as a whole.

This accelerating double layer, although stable in one dimension, is unstable in two or three dimensions due to the Rayleigh–Taylor instability.

9.2.2 *Underlying Theory of the RPA in One-Dimension*

Here, we present the theory of RPA of ultra-thin targets. In the RPA, the thickness of the foil is such that the electrons pushed by the ponderomotive force of the circularly polarized laser light reach the rear end of the target, where they are stopped by the static electric field of the ions, as suggested first in 1D simulations by [16]. Thus the target forms a self-organized double layer. The whole foil, now in the form of a polarized and overdensed plasma, is accelerated by the radiation pressure of the laser. A theory for this accelerated plasma double layer was developed by us [13, 15, 29] to explain the origin of accelerated ions first seen in numerical simulations [16].

9.2.2.1 Effects of Circular Polarization on the Stability and Efficiency of RPA

For efficient acceleration of an ultrathin target, circularly polarized laser is essential. With a linearly polarized laser, electrons move along the radiation electric field and can gain large transverse momentum and be heated; thus consequently the acceleration of protons along the laser propagation direction becomes less efficient. On the other hand, circularly polarized laser has only zero frequency components of ponderomotive force and suppresses the second harmonic component, driving the foil as a light sail. Therefore, the stability and efficiency of RPA is increased with circularly polarized laser.

9.2.2.2 Optimal Thickness of the Ultra-thin Target for Light Sail and the Formation of a Self-Organized Double Layer

During the RPA of an ultrathin foil, the laser ponderomotive force sweeps all electrons in the foil forward until the electrostatic force on the electrons due to the ions left behind balances the ponderomotive force on the electrons at a distance D. These electrons form a charged layer and their electrostatic force now accelerates the ions left behind. When the thickness of the target, Δ, is equal to this distance of maximum charge separation, we obtain optimal thickness $D = \Delta_s$, at which the electrons are pushed to the rear end of the target and the space charge force balances the ponderomotive force $eE_s = F_P(\Delta_s)$ on the electrons. In the large amplitude limit of the normalized laser amplitude $a_0 = e|E|/(m\omega c) >> 1$, we obtain the thickness

$$\Delta_s \cong \frac{4\pi}{\lambda_L}\left(\frac{c}{\omega_p}\right)^2 a_0 = \frac{\lambda_L}{\pi}\left(\frac{\omega}{\omega_p}\right)^2 a_0 \tag{9.1}$$

To avoid wave tunneling, it is required that $\Delta_s > c/\omega_p$ or $2a_0\omega/\omega_p > 1$. For $a_0 = 5$ and $\omega_p^2/\omega^2 = 10$, the optimum foil thickness is $D = \Delta_s = 0.16\lambda_L$, which compares well with the value of $D = 0.2\lambda_L$ considered in the simulation of Yan et al. [16]. In their simulation, they found that with the target thickness $D = 0.2\lambda_L$, the acceleration produces protons with maximum energy gain and minimum energy spread.

If the thickness of the thin foil, D, is equal to the optimal thickness, a compressed electron layer will be formed at the rear surface, whose electric field will attract the ions to within its skin depth. These ions will be pulled back by the inertial force of their acceleration. Together with the electron sheath, they form a double layer, which is accelerated by the radiation pressure. The ions experience an inertial force—$m\mathrm{d}V/\mathrm{d}t$ due to the acceleration of the target by the ponderomotive force, which tends to confine them by reflecting them back to the rear surface. These ions can now be stably trapped in the well formed by the combined forces of the electron sheath and the inertial force of the accelerated frame. It is clear that target thickness $D = \Delta_s$ where electron experiences force balance is important for the double layer formation. This corresponds to the maximum electrostatic field for ion accelerations considered by [16].

9.2.2.3 Light Sail of the Target as a Whole

The electron–ion double layer is accelerated as a whole by the laser radiation pressure. The time development of the energy of the accelerated protons in terms of the laser parameters can be obtained from simple arguments based on energy conservation [13]. As the ion–electron double layer is accelerated by the radiation force, the laser reflectivity $|R|^2$ becomes less than 1 and the radiation force on the double layer per unit area becomes $F = (I_0/c)(1 + |R|^2)$, where I_o is the incident laser intensity. If the velocity of the double layer is V_f, the work done by the

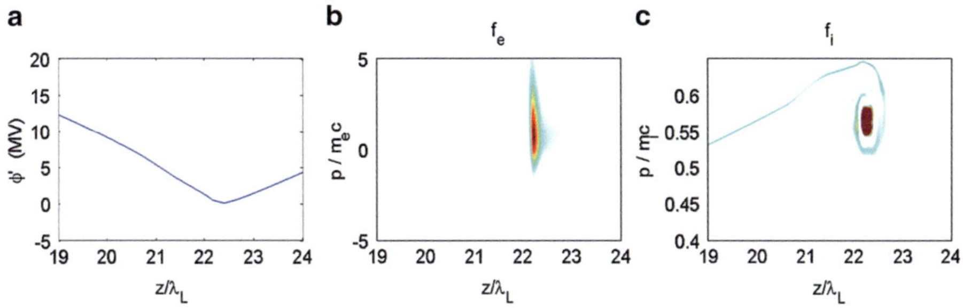

Fig. 9.2 The effective potential (**a**), the electron (**b**) and ion (**c**) phase space density distribution function for target thickness $= 0.2\lambda$, illustrating the importance of trapping ions in the thin double layer to obtain monoenergetic ions. [7]

ponderomotive force per unit area per second is FV_f and we can obtain the radiation force as

$$F = (2I_0/c)(1 - V_f/c)/(1 + V_f/c) \tag{9.2}$$

The equation of motion of the foil is

$$\frac{\mathrm{d}(\gamma_f V_f)}{\mathrm{d}t} = \frac{2I_0}{m_i n_0 \Delta_s c} \frac{1 - V_f/c}{1 + V_f/c} \tag{9.3}$$

where $\gamma_f = (1 - V_f^2/c^2)^{-1/2}$ and $m_i n_0 \Delta_s$ is the mass of the double layer. Equation (9.3) has been analytically solved by [15] and others to obtain ion energy:

$$\varepsilon_i(t) = (\gamma_f - 1)m_i c^2 = ((1 - V_f^2/c^2)^{-1/2} - 1)m_i c^2 \tag{9.4}$$

In the limit of $t/T_L << 1$, the ion energy varies with time as $\varepsilon_i(t) \propto t$, and in the limit $t/T_L >> 1$, it varies as $\varepsilon_i(t) \propto t^{1/3}$. The time for acceleration is limited by the laser pulse length or the duration for the double layer to maintain its stable overdense properties.

Eliasson et al. [7] performed one-dimensional Vlasov simulations and validated the above theory of RPA of ultrathin foil. Figure 9.2 illustrates the potential trap (left panel) in the form a self-organized double layer from the 1D Vlasov simulation. The trapping of ions and formation of double layer can be inferred from the electron and ion phase space density distributions. The theoretical analysis shows that the critical parameter for double layer-trapped monoenergetic ion acceleration is the thickness D of the thin foil. With the laser amplitude $a_0 = 5$, and $\omega_p^2/\omega^2 = 10$, the optimum foil thickness is $\Delta \sim 0.2\lambda_L$ [7]. The stable trapping of ions with optimal thickness is shown in Fig. 9.2. The ion energy evolution given by (9.3) and (9.4) agrees well with the results of 1D PIC simulations [16] for similar parameters. The ion energy increases nearly linearly with time up to $t/T_L \sim 100$ and later increases slowly as $t^{1/3}$.

9.2.3 Rayleigh–Taylor Instability-Induced Transparency and Particle Energy Spectrum Broadening

The Rayleigh–Taylor instability (RTI) of an accelerating thin foil, particularly the nonlinear stage of its development can limit the acceleration achieved by RPA and undesirably broaden the proton beam's energy spectrum so that fewer protons carry the desired energy.

The RTI of a thin foil has recently been studied by Pegararo and Bulanov [35]. It is one of the most important instabilities when a heavy fluid is supported by lighter fluid, arising due to the interchange of the fluids. In the case of RPA, the lighter fluid is the laser radiation and heavy fluid is the overdensed plasma foil. The usual nonlinear development of the RTI is to generate fingers of high and low densities at the most unstable mode with wavelength comparable to the laser wavelength. In the nonrelativistic limit, the fastest growing mode ($x_1 \propto \exp(t/\tau)$) has wavelength l comparable to the laser wavelength and growth rate $\sqrt{g/\lambda}$, where g is the acceleration $\sim 10^{20}\,\mathrm{m/s^2}$. For ultrarelativistic limit, the growth of the amplitude of perturbation is $x_1 \propto \exp(\sqrt[3]{t/\tau_r})$ [35].

During the growth of the RTI, small ripples on the surface of the double layer grow exponentially and these fine structures in the transverse direction perpendicular to laser propagation are converted to large amplitude perturbations with wider transverse periodical structures, in the form of density blobs. During the development of the RTI, the filamentation of the plasma density will produce regions of low-density plasma, transparent to laser radiation. The laser can no longer be reflected effectively by the plasma mirror, and the acceleration stops.

The 2D PIC simulations by [30] confirm that the RTI can destroy the electron layer and widen the energy spectrum of ions. For effective acceleration and production of monoenergetic particles, it is important to maintain the target density as overdense (density above the critical density). During the acceleration, the RTI develops and forms interleaving high-density blobs and low-density regions as shown in Fig. 9.3b, f. The density difference between the high-and low-density regions grows very rapidly and filaments, or fingerlike structures, can be further developed. Once the density of the low-density region of the foil is below critical density (evolution shown in 1st column of Fig. 9.3), the laser light can penetrate that region (evolution shown in 3rd column of Fig. 9.3), the target will be heated, and the particle energy spread will be significantly increased (evolution shown in 4th column of Fig. 9.3). This leads to a leakage of radiation through the target by self-induced transparency, preventing the production and maintenance of monoenergetic protons.

9.2.4 Energy Scaling of RPA Generation of Quasi-Monoenergetic Protons

A direct consequence of the RTI-induced transparency and the broadening of the proton energy spectra is the limitation on the achievable quasi-monoenergetic proton energy or brightness of the particle beam for a given laser power. By defining

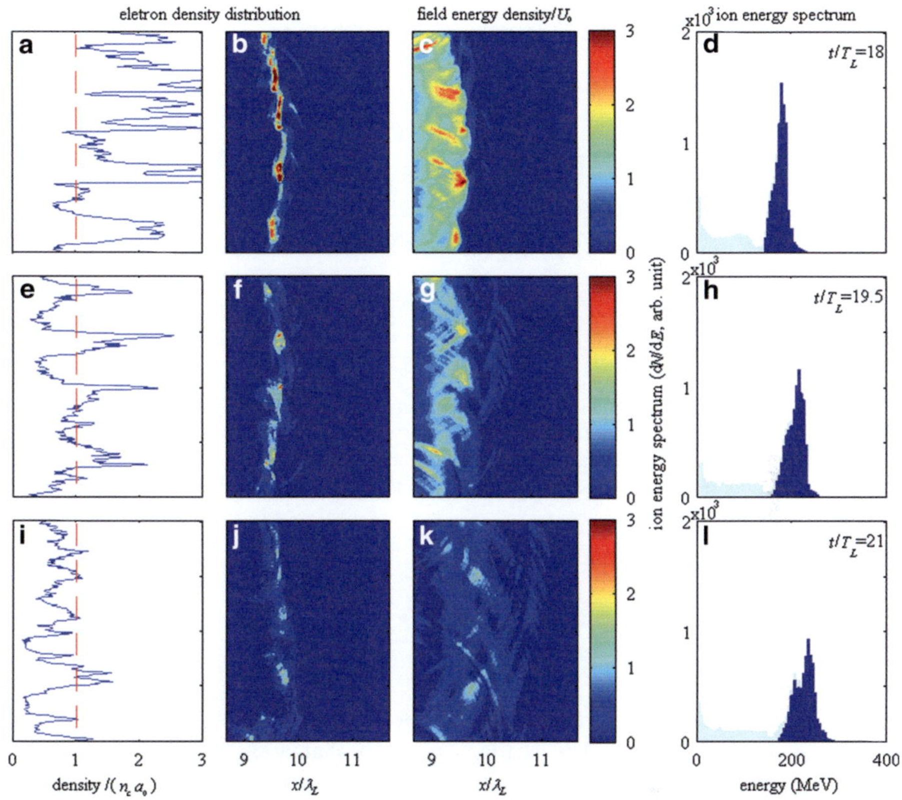

Fig. 9.3 2D PIC simulation of the evolution of averaged Electron density transverse to laser beam (panels in 1st column), Electron density map (panels in 2nd column), electromagnetic field energy density (panels in 3rd column) and energy spectrum of the protons (panels in 4th column) during three instants. *Top*, *middle*, and *bottom rows* are at $t = 18\ T_L$, $19.5\ T_L$, and $21\ T_L$ respectively. The *red* vertical *dashed line* in panels of 1st column shows the critical density value. In the simulation, the domain is $16\lambda_L \times 24\lambda_L$ and divided into $N_x \times N_y = 1024 \times 1536$ grids and the normalized incident laser amplitude is 25, corresponding to 2 petawatt laser power [30]

quasi-monoenergetic particle as particles having energy spread of half-maximum particle flux within 20% of the energy of peak flux, [30] studied the scaling of quasi-monoenergetic proton energy vs normalized incident laser amplitude (Fig. 9.4). In this study, it was found that the foil density changes linearly with incident laser amplitude when the foil thickness was chosen to be the optimal thickness (9.1). The attainable quasi-monoenergetic ion energy is

$$E_{\text{non-rel}} \approx 2\left(\pi \frac{m_e}{m_i}\frac{a_0 t_s}{T_L}\right)^2 m_i c^2 \tag{9.5}$$

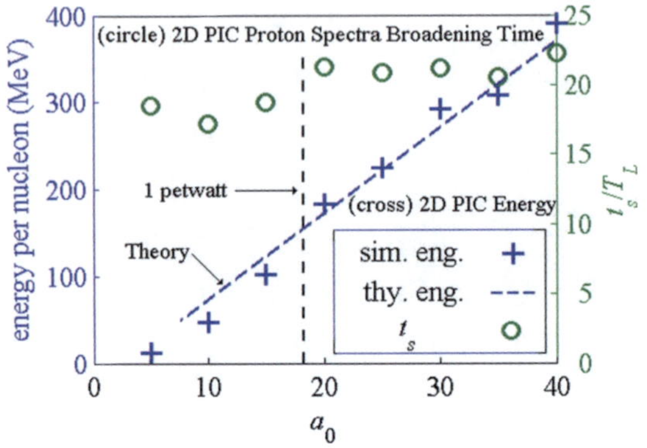

Fig. 9.4 Scaling of quasi-monoenergetic proton energy vs. normalized incident laser amplitude as obtained from PIC simulation (*blue dot*) and theory (*blue dashed curve*). *Green dots* indicate the time when the proton spectra start to broaden and deviate from our definition of quasi-monoenergetic. For normalized laser amplitude ~ 15, the laser power is around 1 PetaWatt [30]

for $\pi a_0 t_s / T_L << m_i / m_e$, and

$$E_{\text{non-rel}} \approx \left(\frac{3\pi}{2} \frac{m_e}{m_i} \frac{a_0 t_s}{T_L} \right)^{1/3} m_i c^2 \tag{9.6}$$

for $\pi a_0 t_s / T_L >> m_i / m_e$, where t_s is the occurrence time of maximum quasi-monoenergetic ion energy. The scaling indicates that with flat ultra-thin targets of optimal thickness, the quasi-monoenergetic proton energy can be as high as 250 MeV for $a_0 = 25$ corresponding to 2 petawatt laser power. Also, the quasi-monoenergetic proton energy is adjustable by changing the incident laser power. However, this scaling is difficult to obtain in real applications since the manufacturing of solid targets made of pure hydrogen is a challenging task. Particle energy that spreads less than 10% of the peak flux energy is also more desirable and sub-petawatt lasers are more realistic for commercialization and for maintaining high repetition rates. Therefore, there are needs for research on laser proton accelerator using practical targets and on suppressing and remediate the RTI.

9.2.5 Promising Experimental Evidence of the RPA of an Ultra-Thin Carbon Target

Due to the stringent requirement on laser quality (high contrast ratio etc.), there have been few experiments to date to demonstrate RPA. Recent experiments by [10] demonstrated that laser RPA of a thin foil might be a promising way of obtaining quasi-monoenergetic ion beams. With 30 TW laser and peak intensity of 5×10^{19} W/cm^2 irradiated on a diamond-like carbon (DLC) foil of thickness 5.3 nm, [10] obtained a quasi-monoenergetic carbon C^{6+} ion beam with a distinct peak energy around 30 MeV and an energy spread ~20 MeV. In comparison with the conventional TNSA scheme with laser irradiation on thick target [18], the conversion efficiency is estimated to increase more than 40 times in the RPA scheme of circularly polarized laser irradiation on ultrathin foil. Although the ion beam energy obtained in this experiment is relatively low and the beam quality is still far from monoenergetic, the experiment does provide promising aspects of RPA of ultrathin targets.

9.3 Other Forms of RPA

9.3.1 Multi-Ion Target Acceleration with RPA

For effective RPA, it might appear most efficient to use the lightest element hydrogen for the target material, as it has minimum mass and therefore is easy to accelerate for a given laser power. However, it is impractical to make pure hydrogen thin targets. On the contrary, multi-ion targets consisting of carbon ions and protons are easily available and can be made ultrathin. In fact the recently reported experiment for quasi-monoenergetic carbon acceleration used diamond-like carbon (DLC) foil [10] of thickness 5.3 nm with about 10% hydrogen impurity. Carbon ions and protons are considered of complementary values for hadron therapy of cancer. The energy requirement for carbon ion therapy is 400 MeV–5 GeV and is 50–250 MeV for proton therapy [31].

For targets with two ion species, it is possible to form self organized triple layers of electrons and two different ion species and the electron back flow to neutralize the heavier layer. Figure 9.5 shows simulation results of the RPA acceleration of an ultra-thin C_2H_4 film. The simulation box is $16\lambda_L \times 24\lambda_L$, and the fully-ionized C_2H_4 foil is placed at $4 < x/\lambda_L$ 4.2 with thickness = 0.2 λ_L. The incident circularly polarized laser is of Gaussian shape with waist size $8\lambda_L$ in the y direction and normalized peak amplitude of the $a_0 = 25$. With carbons providing more electrons and the carbon layer being left behind, it is possible to delay the electron layer from becoming transparent and to accelerate the proton layer by the laser for a longer time, thereby obtaining quasi-monoenergetic protons with higher energy. It can be seen in Fig. 9.5 that the proton layer remains monoenergetic (panel c) while the carbon layer starts to be torn apart due to the RTI. It is expected that with the decrease of hydrogen concentration in the target, the effects of Coulomb explosion due to the repulsion force between the carbon ions and the proton layer becomes important.

9.3.2 RPA of Gas Target

Recently, there has been a great interest in both experiments and modeling of laser proton acceleration with hydrogen gas targets for energetic proton generation [32]. The laser proton acceleration of gas targets is attractive since hydrogen gas is readily

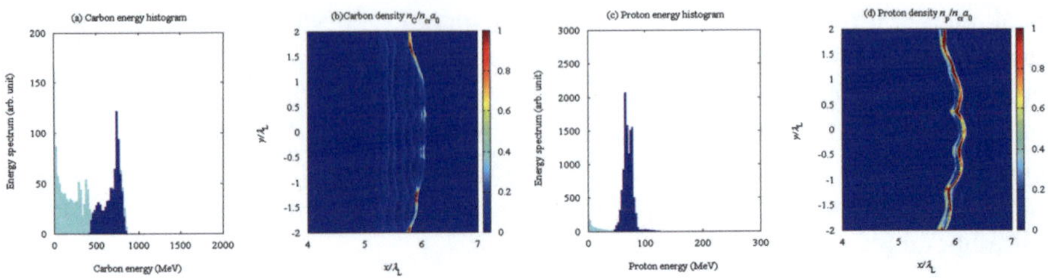

Fig. 9.5 2D PIC simulation of laser-particle acceleration of a C_2H_4 foil. (**a, b**) The Carbon ion energy spectrum and density plot at time $t = 14.5$ laser wave period, respectively; (**c, d**) The corresponding Proton energy spectrum and density plot at the same time instant, respectively. The energy spectrum is recorded within a window of width $2\lambda_L$ in y direction and full extend in x direction for histogram in *cyan color*; and within a window of width $2\lambda_L$ in y direction and $0.5\lambda_L$ width around the location of maximum number of particles in x direction for histogram in *blue color* [30]

available and the laser can be CO_2 laser which is of low cost and long duration and can deliver substantial amounts of energy for proton acceleration.

Here, we present a scheme of laser thin gas target acceleration for quasi-monoenergetic proton generation. The scheme uses gas targets of thickness about several laser wavelengths with the gas density spatial distribution being Gaussian- or square sine-shaped. For thin gas targets (a few laser wavelength), the laser radiation pressure can compress the spatially distributed proton gas into a thin layer of sub-wavelength thickness. The compressed thin layer of proton is over-dense and can therefore be accelerated by the laser RPA until the development of the RTI destroys the opacity of the compressed layer. An example of RPA of a gas target is shown below.

A PIC simulation of laser RPA of a gas target was performed using circularly polarized laser of normalized amplitude $\sim a_0 = 5$ and a hydrogen gas target of thickness $\sim$2.5 laser wavelength with a peak density 20 times the critical density. The PIC simulation domain is within $0 \leq x/\lambda_L \leq 30$ and $0 \leq y/\lambda_L \leq 10$. The gaseous plasma is initially located at $6 \leq x/\lambda_L \leq 8.5$ with initial spatial density profile proportional to $n_0 \sin^2[(x/\lambda_L - 6)\pi/2.5]$ and $n_0 = 20n_c$. The incident circularly polarized laser is a plane wave in transverse direction and the time profile of its amplitude is of 10 T_L in rising, 280 T_L flat, and 10 T_L falling time, respectively. The rising and falling time slope is of square sine shape. Figure 9.6 shows the simulation results of the evolution of the electron, and proton density distributions and the proton energy spectrum.

The simulation demonstrates several key physical processes involved in the laser-thin gas target acceleration, where quasi-monoenergetic protons are obtained. During the first 40–70 laser wave periods, we observed a shock formation. A compressed plasma layer is formed with an enhanced density as high as 60 times the critical density within a region of sub-wavelength scale. The laser radiation pressure is able to accelerate the density peak formed around the shock. This is shown in the first three rows of Fig. 9.6, where the compressed electron-ion layer

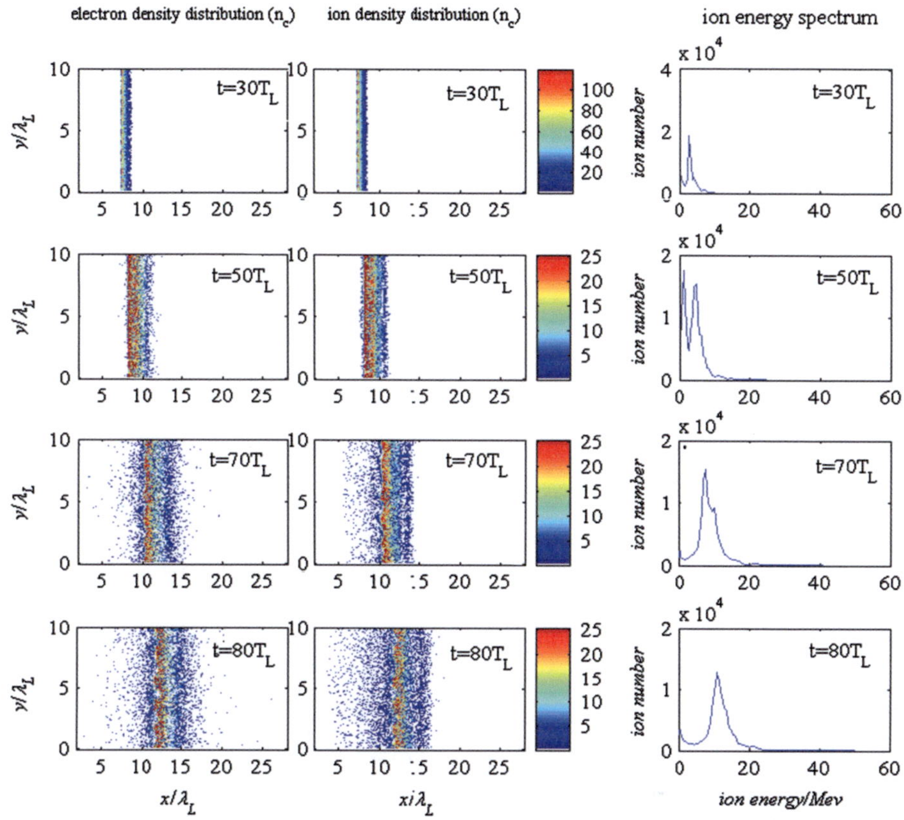

Fig. 9.6 2D PIC simulation results of the evolution of the Electron density distribution (the first column), Ion density distribution (the second column) and Ion energy spectrum (the third column). From the *top row* to the *bottom row*, the time instants are $t = 60\ T_L$, $100\ T_L$, $130\ T_L$, and $160\ T_L$, respectively [36]

travels a distance of $7\lambda_L$ over a time of 130 T_L and the quasi-monoenergetic protons gain energy of about 6 MeV. The quasi-monoenergetic property of the proton beam is finally destroyed due to the RTI and the target becomes broadened and transparent with laser light leaking through the low-density region formed on the target (Fig. 9.6j, k). The proton energy can reach as high as 10 MeV at t $= 160T_L$ (Fig. 9.6l).

Figure 9.7 shows the dependence of the energy of the quasi-monoenergetic protons on the gas target peak density and the incident laser energy in laser-thin gas target acceleration. The simulations use the same spatial profiles of the gas target densities as that in Fig. 9.6. Changing the incident laser amplitude from $a_0 = 5$ to $a_0 = 10$ increases the energy of quasi-monoenergetic proton from about 10 MeV to 16 MeV (see lines a and b in Fig. 9.7). Further increase of the peak gas density from $n_0 = 20\ n_c$ to $n_0 = 30\ n_c$ also helps to further increases the energy of the quasi-monoenergetic protons to 22 MeV (see lines b and c in Fig. 9.7).

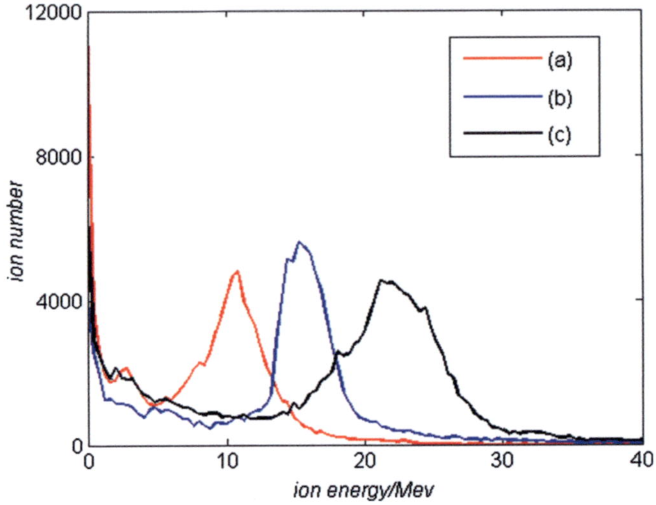

Fig. 9.7 The proton energy spectrum obtained with different incident laser and plasma parameters. The simulation parameter of line (**a**) (*red line*) is normalized laser amplitude $a = 5$, gas target thickness $l_s = 2.5\ \lambda_L$, peak gas target density $n_0 = 20\ n_c$. line (**b**) (*blue line*) is $a = 10$, $l_s = 2.5\lambda_L$, $n_0 = 20\ n_c$, and line (**c**) (*black line*) is $a = 10$, $l_s = 2.5\lambda_L$, $n_0 = 30\ n_c$ [36]

9.4 Medical Implications of Quasi-Monoenergetic Proton Beam Generated from RPA: Laser-Proton Cancer Therapy with the RPA-Based Laser Proton Accelerator

Conventional *cyclotron* or *synchrotron accelerator-based particle therapy* usually requires a particle beam transport system and gantry to steer the particle beam toward the targeted area for treatment. This adds a substantial amount of cost and reduces the flexibility of the conventional proton therapy facility. With the concept of compact laser-proton accelerators, it is feasible that the proton delivery system can have mirrors to steer the laser into the treatment room and the laser strikes onto the target foil right above the patient. Additional subsystems for filtering, monitoring and steering of the proton beam produced from the RPA are needed to make the particle beam ready for patient treatment, but it will be much less demanding in comparison to the conventional systems. It is worth investigating whether systems based on laser-proton accelerators can significantly simplify the overall design as well as reduce the cost.

To assess the feasibility of *laser-proton cancer therapy* with the RPA-based laser-proton accelerator, we need to determine the flux abundance and quality of the proton beam obtained from the laser-proton accelerator and its interactions with human tissues. Therefore, simulations are carried out to model the interaction of protons with water and determine the radiation dosage deposition for particle beams produced from a PIC simulation of laser RPA of ultrathin target.

In the PIC simulation, the incident laser with 1-μm wavelength has an intensity of Gaussian shape with the peak amplitude $a_0 = 15$ along the central axis and the target is a proton target with optimal thickness (9.1). Figure 9.8a shows the overall spatial distribution of the electron and ion density at $t = 17\ T_L$ produced from the PIC simulation as well as the zoom-in view of the proton beam within 2 μm from the center axis. The protons within this window have an energy spectrum shown in Fig. 9.8b with the peak flux at energy 87 MeV and the energy spread of half maximal flux $\sim$ 10% of the peak flux energy. It is seen from Fig. 9.8a that the spatial distribution of energetic protons appears as density blobs with interweaving high and low density regions in y direction and the periodicity is about equal to the laser wavelength. This is due to the development of the RTI as discussed in Sect. 9.2.3 and it contributes to the broadening of the proton energy spectrum (Fig. 9.8b).The proton beam within the sampling window (2 μm around the center axis) is fed to a Monte-Carlo code for radiation dosage calculations resulting from proton-water interactions.

We used the SRIM (the Stopping and Range of Ions in Matter Model, www.SRIM.org) code to calculate the depth and lateral dose distribution of protons energized by laser radiation pressure. SRIM uses a Monte Carlo method to calculate the spatial dosage deposition in matter for protons of a given energy. The overall dosage deposition map from the proton beam is derived by superposing the radiation dosage contributed from each particle fed from the PIC simulation. The results are shown in Fig. 9.9. Figure 9.9a compares the dosage distribution vs. the penetration depth for monoenergetic 87 MeV protons and that for the protons obtained from the PIC simulation (Fig. 9.8b). It can be seen from Fig. 9.9a that protons produced by laser thin foil interaction give high energies "exit" dose and the entrance dose is not too much different from monoenergetic beams. The relatively longitudinal depth width for effective dosage deposition might be useful for limiting the usage of rang modulator in the particle therapy to alter the particle energy and adapt to finite longitudinal size of the tumor.

Next, we estimate the amount of radiation dosage deposition and check whether it is sufficient for the proton therapy. The PIC simulation indicates that quasi-monoenergetic proton beams of peak flux energy $\sim$ 87 MeV with 10% energy spread can be obtained. The corresponding laser power is $\sim$ 1 PW and the energy delivered by the laser to the target is $\sim$ 25 Joule over $\sim$ 17 laser periods. The RPA produces proton beams of energy $\sim$ 0.3 Joule. Within the narrow energy range of 86–90 MeV around the peak flux energy, the total energy of available protons is $\sim$0.06 Joule. The simulations of the interaction of the focused beam with water show that the radiation energy deposited at the dosage peak is about 17% of the total deposited energy, which corresponds to $\sim$ 0.01 Joule particle beam energy deposited at the Bragg peak of the tissue. This amounts to about 20 Gy (Gy = J/kg) radiation dosage deposition, which is sufficient in comparison with the usual requirement of 2 $\sim$ 5 Gy for normal particle radiation therapy.

The energetic protons fed directly from the laser RPA accelerator are distributed within a hole of 4-μm diameter. That transverse motion of the proton beam from the RPA splits the beam into several bunches and the radiation energy deposition also splits into bunches in mm scale (see Fig. 9.9b). Due to the small beam size ($\sim$ 4 μm)

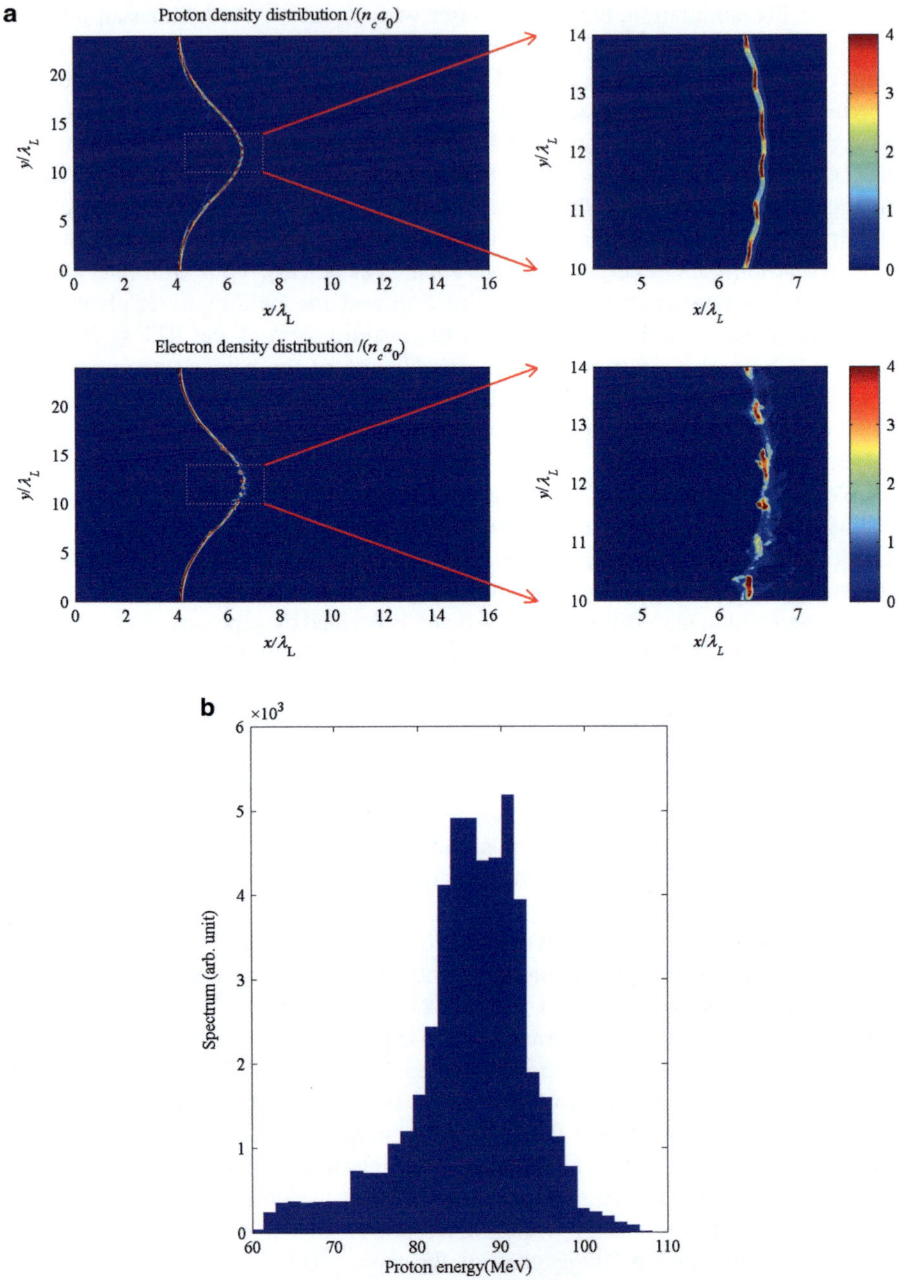

Fig. 9.8 (**a**) *Left panels*: ion and Electron density maps taken at $t = 17$ laser periods from the simulation of laser RPA of an ultra-thin target. Right panels show the zoom-in view of the proton and Electron density within 2 μm from the center, i.e. between 10 and 14 μm along y. (**b**) Energy spectrum obtained for the protons within the 2 μm-radius window around the x axis from the 2D PIC simulation

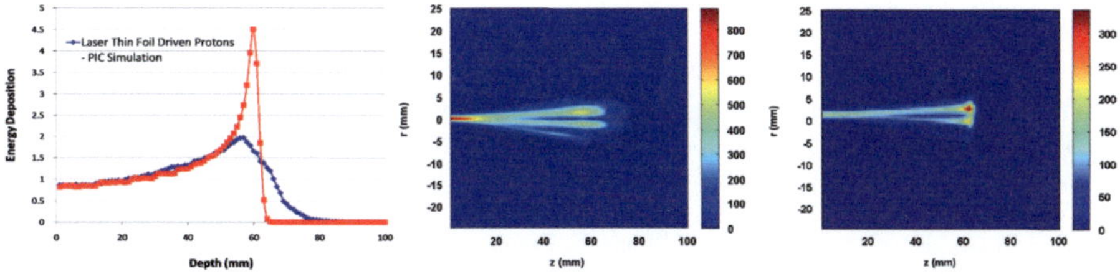

Fig. 9.9 (**a**) *Blue line* is the Energy deposition of protons from the simulation of laser thin foil acceleration. *Red line* is the Energy deposition of same number of monoenergetic 87 MeV protons. (**b**) Energy dosage contour of proton beam entering a water phantom right after it is produced from RPA laser proton accelerator. (**c**) Dose distribution after the proton beam passed through bending magnet and energy selection slit. The selected energy range is from 86 to 90 MeV

at the entrance, the entrance energy dosage is the highest, which will damage the healthy tissue and is not good for proton therapy. Therefore, the proton beam obtained from laser RPA cannot be directly used for proton therapy. Additional systems for filtering, monitoring and steering the proton beam produced from the RPA are needed. For example, by using bending magnet and energy selection slit, the proton beam can be filtered to contain protons with energies from 86 to 90 MeV and the entry diameter is broadened to $\sim$ 2 mm. The profile of radiation energy deposition for this filtered proton beam is shown in Fig. 9.9c and it can be seen that the radiation dosage deposition is greatly improved and more targeted. Further integrated research is needed to investigate the capability, limitation and improvement on laser-proton accelerator and system design for enabling laser proton cancer therapy.

9.5 Summary

In this paper, we have reviewed the physics of quasi-monoenergetic proton generation by RPA and recent advances in the studies of energy scaling of RPA, as well as RPA of multi-ion and gas targets. The scheme for generating quasi-monoenergetic protons with RPA has the potential of leading to table-top accelerators as sources of monoenergetic 50–250 MeV protons. We have also discussed the potential of medical implications, such as particle therapy for cancer treatment, using quasi-monoenergetic proton beam generated from RPA. Although the laser RPA shows promising aspects such as being able to deliver sufficient dosage and simplify system design for applying it in particle therapy, further studies to profile and improve the quality of the particle beam generated from laser RPA and integrate research involving physicists in experiment and radiation therapy are needed. Compact monoenergetic ion sources have also applications in high energy particle physics, space electronics radiation testing, and fast ignition in laser fusion.

Acknowledgements We thank helpful discussions with A. Ting (Navel Research Laboratory) and J. W. Wong (John Hopkins University).

References

1. T. Tajima, J.M. Dawson, Laser electron accelerator. Phys. Rev. Lett. **43**, 267–70 (1979)
2. J. Faure, Y. Glinec, A. Pukhov, S. Kiselev, S. Gordienko, E. Lefebvre, J.P. Rousseau, F. Burgy, V. Malka, A laserplasma accelerator producing monoenergetic electron beams Nature **431**, 541–4 (2004)
3. C.G.R. Geddes, C.S. Toth, J. van Tilborg, E. Esarey, C.B. Schroeder, D. Bruhwiler, C. Nieter, J. Cary, W.P. Leemans,High-quality electron beams from a laser wakefield accelerator using plasma-channel guiding Nature **431**, 538–41 (2004)
4. S.P.D. Mangles, C.D. Murphy, Z. Najmudin, A.G.R. Thomas, J.L. Collier, A.E. Dangor, E.J. Divall, P.S. Foster, J.G. Gallacher, C.J. Hooker, D.A. Jaroszynski, A.J.Y. Langle, W.B. Mori, P.A. Norreys, F.S. Tsung, R. Viskup, B.R. Walton, K. Krushelnick, Monoenergetic beams of relativistic electrons from intense laser plasma interactions Nature **431**, 535–8 (2004)
5. S. Kneip, S.R. Nagel, S.F. Martins, S.P.D. Mangles, C. Bellei, O. Chekhlov, R.J. Clarke, N. Delerue, E.J. Divall, G. Doucas, K. Ertel, F. Fiuza, R. Fonseca, P. Foster, S.J. Hawkes, C.J. Hooker, K. Krushelnick, W.B. Mori, C.A.J. Palmer, K. Ta Phuoc, P.P. Rajeev, J. Schreiber, M.J.V. Streeter, D. Urner, J. Vieira, L.O. Silva, Z. Najmudin, Near-GeV acceleration of electrons by a nonlinear plasma wave driven by a self-guided laser pulse Phys. Rev. Lett. **103**, 035002 (2009)
6. W.P. Leemans, B. Nagler, A.J. Gonsalves, C.S. Tóth, K. Nakamura, C.G.R. Geddes, E. Esarey, C.B. Schroeder, S.M. Hooker, GeV electron beams from a centimetre-scale accelerator Nat. Phys. **2**, 696–9 (2006)
7. B. Eliasson, C.S. Liu, X. Shao, R.Z. Sagdeev, P.K. Shukla, Laser acceleration of monoenergetic protons via a double layer emerging from an ultra-thin foil New J. Phys. **11** 073006 (2009)
8. T. Esirkepov, M. Borghesi, S.V. Bulanov, G. Mourou, T. Tajima, Highly efficient relativistic-ion generation in the laser-piston regime Phys. Rev. Lett. **92**, 175003 (2004)
9. A.A. Gonoskov, A.V. Korzhimanov, V.I. Eremin, A.V. Kim, A.M. Sergeev, Multicascade proton acceleration by a superintense laser pulse in the regime of relativistically induced slab transparency. Phys. Rev. Lett **102** 184801 (2009)
10. A. Henig, S. Steinke, M. Schnürer, T. Sokollik, R. Hörlein, D. Kiefer, D. Jung, J. Schreiber, B.M. Hegelich, X.Q. Yan, T. Tajima, P.V. Nickles, W. Sandner, D. Habs, Radiation pressure acceleration of ion beams driven by circularly polarized laser pulses. Phys. Rev. Lett. **103**, 245003 (2009)
11. O. Klimo, J. Psikal, J. Limpouch, V.T. Tikhonchuk, Monoenergetic ion beams from ultrathin foilsirradiated by ultrahigh-contrast circularly polarized laser pulses Phys. Rev. ST Accel. Beams **1**(1), 031301 (2008)
12. T.V. Liseykina, M. Borghesi, A. Macchi, S. Tuveri, Radiation pressure acceleration by ultraintense laser pulses. Plasma Phys. Contr. Fusion **50**, 124033 (2008)
13. C.S. Liu, V.K. Tripathi, and X. Shao, Laser Acceleration of Monoenergetic Protons Trapped in Moving Double Layer from Thin Foil, ed. by P.K. Shukla, B. Eliasson, L. Stenflo. *Frontiers in Modern Plasma Physics: 2008 ICTP International Workshop. AIP Conference Proceedings* vol 1061, pp. 246–254, 2008
14. A.P.L. Robinson, M. Zepf, S. Kar, R.G. Evans, C. Bellei, Radiation pressure acceleration of thin foils with circularly polarized laser pulses. New J. Phys. **10**, 013021 (2009)
15. V.K. Tripathi, C.S. Liu, X. Shao, B. Eliasson, R.Z. Sagdeev, Laser acceleration of monoenergetic protons in a self-organized double layer from thin foil. Plasma Phys. Contr. Fusion **51**, 024014 (2009)
16. X.Q. Yan, C. Lin, Z.M. Sheng, Generating high-current monoenergetic proton beams by a circularly polarized laser pulse in the phase-stable acceleration regime Phys. Rev. Lett. **100**, 135003 (2008)
17. J. Fuchs, C.A. Cecchetti, M. Borghesi, T. Grismayer, E. d'Humières, P. Antici, S. Atzeni, P. Mora, A. Pipahl, L. Romagnani, A. Schiavi, Y. Sentoku, T. Toncian, P. Audebert, O. Willi, Laser-foil acceleration of high-energy protons in small-scale plasma gradients Phys. Rev. Lett. **99** 015002 (2007)

18. B.M. Hegelich, B.J. Albright, J. Cobble, K. Flippo, S. Letzring, M. Paffett, H. Ruhl, J. Schreiber, R.K. Schulze and J.C. Fernandez, Laser acceleration of quasi-monoenergetic MeV ion beams Nature **439**, 441–4 (2006)
19. P. Mora, Laser driven ion acceleration. AIP Conf. Proc. **920**, 98–117 (2007)
20. A. Pukhov, Three-dimensional simulations of ion acceleration from a foil irradiated by a short-pulse laser. Phys. Rev. Lett. **86**, 3562 (2001)
21. L. Robson, P.T. Simpson, R.J. Clarke, K.W.D. Ledingham, F. Lindau, O. Lundh, T. McCanny, P. Mora, D. Neely, C.G. Wahlström, M. Zepf, P. McKenna, Scaling of proton acceleration driven by petawatt–laser–plasma interactions Nat. Phys. **3** 58–62 (2007)
22. H. Schwoever, S. Pfotenhauer, O. Jäckel, K.U. Amthor, B. Liesfeld, W. Ziegler, R. Sauerbrey, K.W.D. Ledingham, T. Esirkepov, Laser-plasma acceleration of quasi-monoenergetic protons from microstructured targets Nature **439**, 445–8 (2006)
23. S. Ter-Avetisyan, M. Schnürer, P.V. Nickles, M. Kalashnikov, E. Risse, T. Sokollik, W. Sandner, A. Andreev, V. Tikhonchuk, Quasimonoenergetic deuteron bursts produced by ultraintense laser pulses. Phys. Rev. Lett. **96**, 145006 (2006)
24. S.C. Wilks, A.B. Langdon, T.E. Cowan, M. Roth, M. Singh, S. Hatchett, M.H. Key, D. Pennington, A. MacKinnon, R.A. Snavely, Energetic proton generation in ultra-intense lasersolid interactions Phys. Plasmas **8**, 542–9 (2001)
25. L. Yin, B.J. Albright, B.M. Hegelich, K.J. Bowers, K.A. Flippo, T.J.T. Kwan, J.C. Fernández, Monoenergetic and GeV ion acceleration from the laser breakout afterburner using ultrathin foil Phys. Plasmas **14**, 056706 (2007)
26. M. Chen, A. Pukhov, T.P. Yu, Z.M. Sheng, Enhanced collimated GeV monoenergetic ion acceleration from a shaped foil target irradiated by a circularly polarized laser pulse. Phys. Rev. Lett. **103**, 024801 (2009)
27. B. Qiao, M. Zepf, M. Borghesi, M. Geissler, Stable GeV ion-beam acceleration from thin foils by circularly polarized laser pulses Phys. Rev. Lett. **102**, 145002 (2009)
28. B. Qiao, M. Zepf, M. Borghesi, B. Dromey, M. Geissler, A. Karmakar, P. Gibbon, Radiation-pressure acceleration of ion beams from nanofoil targets: the leaky light-sail regime Phys. Rev. Lett. **105**, 155002 (2010)
29. C.S. Liu, X. Shao, B. Eliasson, T.C. Liu, V. Tripathi, G. Dudnikova, R.Z. Sagdeev, Laser acceleration of quasi-monoenergetic protons via radiation pressure driven thin foil, in modern challenges in nonlinear plasma physics. AIP Conf. Proc. **1320**, 104–110 (2011a)
30. T.C. Liu, X. Shao, C.S. Liu, J.J. Su, B. Eliasson, V.K. Tripathi, G. Dudnikova, R.Z. Sagdeev, Energetics and energy scaling of quasi-monoenergetic ions in laser radiation pressure acceleration, Phys. Plasmas **18**, 123105 (2011)
31. B. Jones, The case for particle therapy Br. J. Radiol. **78,** 1–8 (2005)
32. A.J. Palmer Charlotte, N.P. Dover, I. Pogorelsky, M. Babzien, G.I. Dudnikova, M. Ispiriyan, M.N. Polyanskiy, J. Schreiber, P. Shkolnikov, V. Yakimenko, Z. Najmudin, Monoenergetic proton beams accelerated by a radiation pressure driven shock Phys. Rev. Lett. **106**, 014801 (2011)
33. K.W.D. Ledingham, W. Galster, R. Sauerbrey, Laser-driven proton oncology—a unique new cancer therapy?. Br. J. Radiol. **80**, 855–8 (2007)
34. C.M. Ma, I. Veltchev, E. Fourkal, J.S. Li, W. Luo, J. Fan, T. Lin, A. Pollack, Development of a laser-driven proton accelerator for cancer therapy. Laser Phys. **16**, 639–646 (2006)
35. F. Pegoraro, S.V. Bulanov, Photon bubbles and ion acceleration in a plasma dominated the radiation pressure of an electromagnetic pulse. Phys. Rev. Lett. **99**, 065002 (2007)
36. M.Q. He, X. Shao, C.S. Liu, T.C. Liu, J.J. Su, G. Dudnikova, R.Z. Sagdeev, Z.M. Sheng, Quasi-monoenergetic Protons Accelerated by Laser Radiation Pressure and Shocks in Thin Gaseous Targets. submitted to Phys. Plasmas (2012)

Index

K. Yamanouchi et al. (eds.), *Progress in Ultrafast Intense Laser Science VIII*,
Springer Series in Chemical Physics 103, DOI 10.1007/978-3-642-28726-8,
© Springer-Verlag Berlin Heidelberg 2012

MIX
Papier aus verantwortungsvollen Quellen
Paper from responsible sources
FSC® C105338

If you have any concerns about our products,
you can contact us on
ProductSafety@springernature.com

In case Publisher is established outside the EU,
the EU authorized representative is:
Springer Nature Customer Service Center GmbH
Europaplatz 3, 69115 Heidelberg, Germany

Printed by Libri Plureos GmbH
in Hamburg, Germany